Anthropology and Dialectical Naturalism

A Philosophical Manifesto

Brian Morris

BLACK
ROSE
BOOKS

Montréal/Chicago/London

Black Rose Books No. UU419

Title: Anthropology and dialectical naturalism : a philosophical manifesto / Brian Morris.
Names: Morris, Brian, 1936- author.
Identifiers: Canadiana (print) 2020020212X | Canadiana (ebook) 20200202146 | ISBN 9781551647449 (hardcover) | ISBN 9781551647425 (softcover) | ISBN 9781551647463 (PDF)
Subjects: LCSH: Anthropology—Philosophy.
Classification: LCC GN33 .M67 2020 | DDC 301.01—dc23

Cover by Tina Carlisi.

C.P.35788 Succ. Léo-Pariseau
Montréal, QC, H2X 0A4
CANADA
Explore our books and subscribe to our newsletter:
blackrosebooks.com

Ordering Information

CANADA	USA/INTERNATIONAL	UK/IRELAND
University of Toronto Press	University of Chicago Press	Central Books
5201 Dufferin Street	Chicago Distribution Center	50 Freshwater Road
Toronto, ON	11030 South Langley Avenue	Chadwell Heath, London
M3H 5T8	Chicago, IL 60628	RM8 1RX
1-800-565-9523	(800) 621-2736 (USA)	+44 (0) 20 8525 8800
utpbooks@utpress.utoronto.ca	(773) 702-7000 (International) orders@press.uchicago.edu	contactus@centralbooks.com

Table of Contents

To the memory of
Stephen Nugent (1950-2018),
a life-long colleague
and radical scholar.

PREFACE

For more than forty years I have been teaching anthropology at Goldsmiths College, University of London, and engaging in ethnographic research studies in both India (among hunter-gatherers), and in Malawi (among peasant smallholders). During that time anthropologists have adopted—sometimes with remarkable enthusiasm—a wide variety of theoretical paradigms: structural-functionalism, structuralism (especially that associated with Levi-Strauss), structural Marxism, transactional analysis, symbolic anthropology, postmodernism, sociobiology, network theory, and the latest theory [sic] to fill the anthropological journals, namely, animistic science (which is a complete oxymoron!).

Given my own background, I never embraced any of these illustrious theoretical paradigms. Indeed, I have never considered myself a *real* anthropologist, even though I was initiated into anthropology at one of its most famous shrines. In fact, I can claim to be a direct intellectual descendent, via Edmund Leach and James Woodburn, of Bronislaw Malinowski, one of the founders of British Social Anthropology. For I have always felt that I belonged to a tribe of scholars that went extinct in the nineteenth century: they described themselves as naturalists, as students of natural history. My early intellectual heroes were therefore not anthropologists, nor philosophers, not even academics: they were people like Charles Darwin, Richard Jefferies, Peter Kropotkin, Ernest Thompson Seton, Jean-Henri Fabre, Frances Pitt, Seton Gordon, and W.H. Hudson.

My intellectual tendencies and aspirations have therefore always been that of an evolutionary naturalist, fundamentally realist and historical, with a strong ecological sensibility. This present text, almost inevitably, runs completely "against the current" of much contemporary anthropological theory. It offers, as a manifesto, both an outline and the advocacy of dialectical (evolutionary) naturalism (emergent materialism) as the most viable and coherent metaphysic to complement and to ground anthropology as a historical social science.

As in my earlier writings, I have tried to keep this manifesto free of academic pretensions and academic jargon, and to write in a style that is lucid and readable. It will I trust appeal to scholars, students, and radical activists, as well as to the general reader.

Again, I should like to express my thanks to my friend Sheila Camfield for kindly typing up the manuscript.

BRIAN MORRIS

May 15th, 2017

INTRODUCTION

Moving against the current

This book does not offer you a completely new way of understanding human social life, let alone "reality"; nor does it seek to challenge the very foundations—no less!—of anthropology. The philosophical pretensions of some academic scholars are indeed quite mind-boggling, especially those with a penchant for Oriental mysticism or esoteric shamanic cults (Capra 1982; Kohn 2013). My aim is much more modest, and is in no way original—apart from the fact that, like all serious scholarship, there is a creative dimension to my own writings and to the present manifesto.

It is not original because I firmly situate myself in a metaphysical and ethical tradition that goes back more than two thousand years, to the earliest materialist sages of Asian philosophy—Kapila, Lao Tzu and the Buddha—and to the Latin scholar Lucretius, whose classic poem "*De Rerum Natura*" (on the nature of things) gave us a splendid account of the philosophical naturalism of the Greek scholar Epicurus (Morris 1990A; O'Keefe 2010). This philosophical tradition was quite distinct from that of the clerics, theologians, and idealist philosophers like Plato and Sankara. It was a tradition that was later expressed by Spinoza and the radical Enlightenment of the eighteenth century, and eventually came to be developed and historicized in the evolutionary naturalism of such seminal figures as Ludwig Feuerbach, Charles Darwin, Karl Marx, Alexander Von Humboldt, George Henry Lewes, Herbert Spencer, Peter Kropotkin, and Roy Wood Sellars.

The aim of this manifesto is not only to make explicit my own metaphysical worldview—which has long formed the background to my own ethnographic studies in Malawi, specifically on the relationship between people and land (and its biota) (Morris 2016)—but also to advocate evolutionary (dialectical) naturalism as the most valid and viable philosophical basis or grounding for my conception of anthropology as a historical socio-ecological science.

I naturally offer my own "take" or interpretation of evolutionary naturalism, but to an important degree I consider historical materialism (Karl Marx), emergent materialism (Mario Bunge), cultural materialism (Marvin Harris), and dialectical naturalism (Murray Bookchin) to be virtual synonyms of Roy Wood Sellars' (and Darwin's) evolutionary naturalism, at least in a broad *ontological* sense, even if each of these scholars articulated quite distinctive epistemologies and had contrasting ethical and political concerns.

This book is very much "against the current", as the Buddhists would say, for its essential thesis—the advocacy of evolutionary (dialectical) naturalism—runs completely counter to broad swathes of recent anthropological "theory". I recognize, of course, the diversity within anthropology—that many scholars continue to reaffirm a form of philosophical materialism (Murphy and Margulis 1995; Lett 1997; Kuznar 1997; Hann and Hart 2011), and, in following Heidegger, many ethnographers privilege the concept of "ontology", thereby implicitly suggesting a materialist metaphysic. Even so, many ethnographers, as ardent social constructionists, invariably continue to view the world through the lens of idealist philosophy.

However, this manifesto is not only a response to and a rebuttal of the extremes of postmodernism (cultural idealism) and sociobiology (reductive materialism) but also of much currently fashionable "theory" (namely perspectivism, relational epistemology, and network theory) that has been grouped under the label "animistic science"—a complete oxymoron on par with Christian (creation) science. Embracing (it appears?) neo-paganism (tribal animism as a religious metaphysic, and an ancient form of idealism), some anthropologists seem intent on either turning anthropology into an adjunct of religious studies or into a shamanic cult. Lacking any sense of history, some anthropologists (e.g., Scott 2013) have a mindset that hardly goes beyond the eighteenth-century Enlightenment and views the only alternative to Cartesian dualism and mechanistic science (itself a form of theistic animism) as being the revival and reaffirmation of another religious worldview—namely tribal animism (spiritualism). These animist anthropologists do not seem to have heard of Darwin, Marx and Engels, or of Michael Bakunin, or of George Henry Lewes and Roy Wood Sellars and evolutionary materialism, let alone social ecology (Morris 2012).

The notion that an earlier generation of anthropologists, because they focussed on human social life and may have lacked an ecological sensibility, were thereby Cartesian dualists is quite misleading, if not completely facile. Equally facile is the equation of "modernity" (so called) with "naturalism" (Descola 1996)—failing to recognize that Christian animism has long been an integral aspect, along with the capitalist economy and the modern nation-state, of both "modernity" and the colonial "encounter", and that empirical naturalism is an intrinsic part of all human societies, including both hunter-gatherers and peasant smallholders.

Many anthropologists have revamped the concept of animism to mean not a belief in spiritual beings, as Edward Tyler (1871) defined it, but to imply a recognition that the world—specifically the biosphere—is animate, comprised of living beings. By such criteria who, apart from diehard Cartesian mechanistic philosophers, is not an animist?!

This manifesto then differs from the prevailing currents of thought in anthropology in essentially four ways:

Firstly, whereas many anthropologists, for example Marshall Sahlins and Tim Ingold, tend to dismiss the philosophical debates between the respective merits of idealism (whatever variety) and materialism as "weary", "irrelevant", or simply a waste of time (Sahlins 1976; Palsson 1996), I affirm the importance of articulating a philosophical perspective. Indeed, I strongly emphasize the importance of embracing a form of evolutionary naturalism as the most relevant and viable grounding for any coherent realist ethnography or anthropological theory (on the advocacy of a realist anthropology, see Zeitlyn and Just 2014).

We have long been subjected to the rhetoric of being "opposed" to "metaphysics" by both Heidegger and the logical positivists, who of course articulated their own brand of metaphysics (respectively mystical idealism and positivism or phenomenalism). But it has to be recognized that all scholars, including anthropologists, express, if only implicitly, some philosophical worldview (metaphysics)—some theory about the nature of the world (reality) and the place of humans within it. Thus, all anthropologists, even though focussing on epistemology, implicitly express some philosophical worldview (or they simply engage in metaphysical ramblings endlessly quoting Heidegger). As I detail in this manifesto, this has often entailed either the embrace of the Neo-Kantian cultural idealism (social constructionism) or of positivism, whether described as radical empiricism, ethnomethodology, network theory, existential anthropology, or empirical realism.

The postmodernist's claim that ontological (metaphysical) reasoning is inherently oppressive is quite facile. It is a cover for their own cultural idealism, a philosophy that tends to deny the fact that humans are fundamentally earthly beings, and it verges on nihilism.

Secondly, whereas many anthropologists, following in the wake of critical theorists and postmodernist philosophers, denigrate or even completely reject the Enlightenment as a "metaphysical illusion" (Palsson 1996: 77), misleadingly holding it responsible for the horrors of the twentieth century (two world wars, the Holocaust, and the ecological crisis). I firmly defend and uphold the Enlightenment. Indeed, I highlight the gross *misrepresentation* of the Enlightenment by its postmodernist critics, and emphasize the importance of reaffirming this intellectual tradition. For in its advocacy of historical naturalism as an ontology; in its promotion of science and reason, thus upholding a universalist, ratio-empiricist epistemology; and, finally, in its affirmation of such universal values as freedom and equality, human social solidarity, free and disinterested enquiry, religious tolerance, and a humanist (as opposed to a theological) form of ethics, the "spirit" of the radical Enlightenment needs to be acknowledged and reaffirmed—not dismissed as an "illusion" by trendy anthropologists.

Thirdly, whereas many anthropologists, and social scientists more generally, have over the recent decades fervently embraced cultural idealism (social constructionism) or some other form of *anti-realism*, this manifesto strongly affirms both ontological realism—the thesis that the material world pre-existed and is completely independent of human cognition—and an emergent materialist metaphysic. I therefore reject the idea that "nature" or "tigers" or "trees" are human "inventions" or "social constructs", or simply cultural phenomena (Dwyer 1996; Seeland 1997), or that bacteria did not exist until discovered by biologists in the late nineteenth century (Latour 1988). Presumably the epiphytic orchid *Bulbo-phyllum malawiensis* did not exist, according to these anti-realist anthropologists, until I described it in the proceedings of the Linnean Society in 1967!

But, of course, a materialist philosophy, as I shall insist in this manifesto, does not imply that time, the human mind (or that of other animals) or human symbolic culture do not exist, or that these things have no relevance: only that they have no reality independent of material things, whether physical objects, organisms (including human persons), or social systems.

Finally, whereas there has been a common tendency among anthropologists, again following in the wake of such postmodernists as Gilles Deleuze and Richard Rorty, to repudiate both the correspondence theory of (factual) truth and the concept of representation, I firmly—against the current—defend and stress the vital importance of both concepts in developing and promoting anthropology as a viable historical science.

This manifesto, in advocating evolutionary naturalism—a philosophy that combines ontological realism, a dialectical epistemology, and a metaphysics of emergent (systemic) materialism—consists of three parts.

In the first part, *Anthropology, Realism and the Enlightenment*, I provide essential background material to the study. After some initial reflections on my own intellectual trajectory as an anthropologist, I discuss in chapter one the dual heritage of anthropology and the relation of the discipline to what many scholars have described as the essential "paradox" of human life. This relates to the fact that humans, as biological organisms, are essentially an intrinsic *part* of the material world—Spinoza's "nature"—while at the same time, through their self-consciousness, intense forms of sociality, and symbolic culture, they are in important respects somewhat unique in perceiving themselves as *separate* from nature. I conclude this first chapter with a critical discussion of postmodern anthropology, highlighting its limitations given its anti-realist tendency and its attempt to reduce social facts to cultural symbols. I then discuss two important intellectual traditions within European culture, which have been fundamental in the development of philosophical naturalism: namely the Greek legacy, focussing on Aristotle's attempt to bring philosophy "down to earth" (chapter two), and the eighteenth-century Enlightenment (chapter three). As indicated above, I defend the Enlightenment against its many critics, and highlight its intellectual virtues. It has always seemed rather perplexing to me that postmodern philosophers and anthropologists, in their rejection of universal values (humanism) and in their "science bashing" (M. Harris 1995), should, while affirming their radical politics, virtually align themselves with the Counter-Enlightenment and with the basic tenets of fascist ideology. Nazism, as an ideology and practice, in fact represented the complete collapse and rejection of Enlightenment humanism within industrial capitalism. I close part one (chapter four) with a critique of the anti-realist intellectual currents that became so pervasive in anthropology and the social sciences towards the end of the last century, and outline and strongly affirm the salience of ontological realism as an essential grounding for anthropological studies.

In part two, *Varieties of Materialism,* I critically outline the three main forms of philosophical materialism that continue to have contemporary relevance, namely, Darwin's evolutionary naturalism, Marx's dialectical materialism, and Mario Bunge's emergent materialism. As a prelude (chapter five), I discuss philosophical materialism more generally as a worldview, viewing it as synonymous with a non-reductive form of naturalism. In this chapter, I make a plea for a *dialectical* approach in interpreting the relationship between concrete particulars—the material entities that constitute the world—and their properties and relations, as well as the relationship between material things and events.

Darwin's evolutionary materialism is the subject of chapter six, and besides outlining the basic tenets of the naturalist's paradigm, I critically engage with the gene-centred perspective of sociobiology and the various forms of sociobiology that have so attracted anthropologists: evolutionary psychology, the theory of memetics, and gene-culture co-evolution.

I then turn to Marx's dialectical materialism, for Marx explicitly acknowledged that Darwin's evolutionary theory confirmed his own approach to socio-economic life, which has been widely described as the materialist conception of (human) history. In chapter seven I thus discuss Marx and Engels conception of "dialectics" as a mode of thought, the sources of Marx's historical materialism, and the various interpretations of Marx that have emerged within the Marxist tradition. I conclude the chapter with a discussion of the dialectical biology which is particularly associated with Stephen Jay Gould, Richard Lewontin, and Richard Levins. These scholars advocated a dialectical approach to biology and to human social life, in opposition to the reductionist gene-centred approach of such sociobiologists as Edward Wilson and Richard Dawkins.

The emergent materialism of Mario Bunge is outlined in two interrelated chapters. In chapter eight I discuss Bunge's hylorealism: his attempt to combine realism—both ontological and critical (scientific)—with a materialist metaphysic. This is accompanied by a critical assessment of the four forms of anti-realism that were prominent in philosophy and the social sciences for much of the twentieth century, namely, philosophical hermeneutics, Husserl's phenomenology, social constructionism—embraced by many anthropologists—and logical positivism. I then turn, in chapter nine, to Bunge's own conception of systemic philosophy, noting its antecedents in the writings of George Henry Lewes and Roy Wood Sellars, and critically assess the limitations of various alternative approaches—spiritual holism, atomism (methodological individualism), structuralism, and positivism—to emergent materialism.

In the third and final part of the manifesto, *Anthropology as a Historical Science,* I discuss, in three chapters, the various conceptual issues relating to the development and affirmation of anthropology as the science that studies "what it means to be a human being", as Immanuel Kant famously expressed it. It is a science, as I conceive it, that is both humanistic (historical) and empirical (ecological). In chapter ten I therefore strongly defend—against the current—the correspondence theory of truth (accepted by almost everyone apart from idealist philosophers, scientific positivists, and postmodern anthropologists), as well as the salience of the concept of representation. This is particularly relevant with respect to ethnographic—empirical—studies, and I view the postmodernist opposition between evocation and representation to be a false and sterile dichotomy. In chapter eleven, on the dialects of social life, I argue for a dialectical synthesis between the opposing concepts of (methodological) individualism and (cultural) holism, and between social structure and human agency.

In the final chapter of the manifesto (chapter twelve) I make a strong plea to retain the "dual heritage" of anthropology, against the postmodernists (or neo-animists), who would reduce anthropology to a form of literature or cultural poetics (or even autobiography), and against the sociobiologists, who wish to reduce anthropology, and even the humanities more generally, to a branch of biology. I thus stress the need to combine hermeneutics (and phenomenology)—with an emphasis on ethnographic fieldwork and cultural diversity (and understanding)—and empirical science—with an emphasis on human origins, social ecology, and world history, and thus with explanations of human social life and culture.

PART ONE

ANTHROPOLOGY, REALISM, AND THE ENLIGHTENMENT

Chapter One

The Anthropological Background

In this opening chapter, I aim to provide some background material to the manifesto and offer some initial reflections on my own intellectual trajectory. I focus on two topics.

The first is to introduce anthropology as a form of enquiry, particularly in relation to what many scholars have described as the essential "paradox" of human life. This is, namely, that humans are an intrinsic part of the natural world as organisms; yet, through their self-consciousness, intense sociality, and symbolic culture, they are in many ways unique, in having a sense of being from nature.

Secondly, I offer some critical reflections on postmodern anthropology, which became an important tendency within the discipline towards the end of the last century.

Prologue: An Intellectual Itinerary

When I became a student-teacher at Brighton College of Education in the late 1960s, the intellectual scene was dominated by three schools of thought: linguistic philosophy, behaviourist psychology, and, with respect to both sociology and anthropology, structural functionalism.

The inspiration for linguistic philosophy was the enigmatic scholar Ludwig Wittgenstein. He was considered by many to be a kind of oracle. Having abandoned his earlier logical positivism, Wittgenstein viewed philosophy not as a scholarly pursuit concerned with understanding the world and the nature of reality, nor with fostering human wisdom (*sophia*) but rather as concerned purely with conceptual analysis. Philosophy, as the later Wittgenstein advocated, was a kind of "therapy" that attempted to unravel, or rather dissolve all philosophical problems. This could be done, we were told, simply by an appeal to the language "games" of everyday social life. Philosophy only needed to focus, his acolytes stressed, on the analysis of the common-sense usage of language. Small wonder that linguistic philosophy came to be later described as the "greatest tragedy" of philosophy in the twentieth century (Magee 1997: 43).

Fortunately for me, as part of my education at the college, I was prompted to read the important writings of John Dewey and Alfred North Whitehead, both of whom had written classic texts on the aims of education. Significantly, both these scholars, while offering sterling critiques of Cartesian metaphysics, still retained a healthy respect for the natural sciences—unlike later postmodern anthropologists.

Behaviourist psychology is now regarded as a thing of the past—quite outdated. But when I was a student-teacher this form of psychology was dominant in educational circles. I not only attended a lecture by the renowned B.F. Skinner, when he visited the University of Sussex—the auditorium was packed and the psychologist Marie Jahoda provoked him with some very searching questions—but the standard textbook on educational psychology was virtually a primer in the psychology of behaviourism (Stones 1966). Moreover, as part of my education, I was shown several films illustrating Skinner's theory of operant conditioning. They mostly depicted, of course, the behaviour of pigeons and rats in a maze! It is now strange to recall that behaviourist psychology was then viewed as a "Copernican revolution" in our way of thinking about human life, and Skinner heralded as an original thinker on part with the pioneer evolutionary naturalist Charles Darwin (Wheeler 1973: 5; Morris 1991: 117-123).

But fortunately—again!—I also took courses at the college on "child development" and the "education of persons" and thus came to critically engage with, for me then, the illuminating writings of Erik Erikson, Jean Piaget, Margaret Mead, and Gordon Allport.

My main course at Brighton College of Education was sociology, taught in an inspiring way by John Raynor and Tony Atcherley. Sociology was then a rather novel subject. It was looked upon with some disfavour, if not disdain, both by the government and by the college, as it was viewed as a politically subversive discipline. But ironically the scholar who was then viewed as the "guru" of the subject was the American sociologist Talcott Parsons, who was hardly a subversive thinker! His book *The Social System* (1951) was considered by my tutors to be the veritable "Bible" of sociology. For through almost-Herculean labours, Talcott Parsons had achieved a grand synthesis: combining the holistic or social systems approach of Emile Durkheim with the atomistic or social action theory associated with Herbert Spencer, Max Weber, and the classical economists. Tending to postulate a homology between cultural norms and human actions, and recognizing three analytic domains—the cultural, social, and personality systems—Talcott Parsons' theory of social life, known as structural functionalism, was almost universally acclaimed as a tour de force in social theory. I have to admit that when I read *The Social System* as a student-teacher I could not understand a word of it. I became completely lost in its labyrinth of abstruse concepts. Other writings, however, came to my rescue, for I discovered C. Wright Mills' classic text *The Sociological Imagination* (1959). This book offered me a strident critique of Talcott Parsons' "grand theory", describing it as a form of "cloudy obscurantism" that tended to ignore the crucial importance of history, the dynamics of power relations, and the complex relationship between historical societies and the human subject. The book, to me, was like a breath of fresh air.

My misgivings with regard to Talcott Parsons' social theory were further bolstered by stimulating discussion I had in May 1970 with the Jewish scholar Basil Bernstein. Now a largely forgotten figure, Bernstein extolled to me the virtues of a more politically engaged and a more vibrant form of sociology. Deeply influenced by symbolic interactionism and the Chicago School of Sociology—which emphasized both language and detailed ethnographic studies—Bernstein, like Mills was an independent and original thinker (on Bernstein, see Castelnuovo and Kotik-Friedgut 2015; on Wright Mills see Horowitz 1983; Morris 2014A: 171-177).

Neither linguistic philosophy, nor behaviourist psychology, nor Talcott Parson's structuralist-functionalism—all forms of positivism—held much appeal to me, for my intellectual tendencies and aspirations tended to be those of an evolutionary naturalist, fundamentally realist and historical. Indeed, all my early intellectual heroes—with which I avidly read during my youthful years—were neither anthropologists nor philosophers, nor even academics; they were people like Charles Darwin, Richard Jefferies, Ernest Thompson Seton, Jean-Henri Fabre, Frances Pitt, Seton Gordon, and W.H. Hudson. All empirical naturalists, all popular writers, and, apart from Darwin, all long forgotten.

When I was accepted as a postgraduate student in anthropology at the London School of Economics in 1970, linguistic philosophy, behaviourist psychology, and structuralist-functionalism in anthropology were all on the wane. Structural-functionalism, in particular, became the subject of a welter of criticism, even though, under its intellectual auspices, it produced many pioneering ethnographical studies and was embraced by many scholars: A.R. Radcliffe-Brown, Meyer Fortes, Isaac Schapera, Robert Merton, Audrey Richards, and Raymond Firth are especially noteworthy. As a novitiate anthropologist I actually wrote an essay entitled *Fourteen Alternatives to Structural Functionalism,* to be rebuked by my tutor Jean La Fontaine for writing with too broad a brush. Alas! My rambling style—as you see while you read this book—is still an intellectual failing of mine!

In the essay I critically discussed some of the alternatives to structural-functionalism then being developed within anthropology. These included, for example, cultural ecology (Julian Steward), symbolic anthropology (the influence of Wittgenstein on Mary Douglas was clearly evident), cultural materialism (Marvin Harris), world-system theory (Eric Wolf), cultural hermeneutics (Clifford Geertz), network analysis and an engagement with "social dramas" (scholars associated with Max Gluckman), and finally, the methodological individualism of Frederick Barth (see Eriksen and Nielsen 2001: 78-104).

But the approach that seems to have really captured the anthropological imagination around 1970, embraced with enthusiasm by a wide range of my contemporaries, was structuralism—whether the psychoanalytic version of Jacques Lacan, the structural Marxism of Louis Althusser, or the structural anthropology of Claude Levi-Strauss. Each of these scholars seems to have generated a cult-like following within anthropology. Inspired by the structural linguistics of Ferdinand de Saussure, all three scholars had in common a complete disregard for history (historicism), advocating instead a kind of synchronic science, and an aversion to any emphasis on subjective experience, or what came to be known as the philosophy of the subject (humanism). Given the widespread enthusiasm for structuralism—which was largely a reaction against and repudiation of Husserl's phenomenology and Sartre's existentialism (both of which had been highly popular in France at the end of the Second-World War)—the two decades between 1960 and 1980 came in fact to be described as the "age of structuralism" (Kurzweil 1980; Morris 2014A: 609-655).

Alas! Unlike many of my contemporaries, including my late friends Olivia Harris and Madan Sarup, I could not muster any enthusiasm for any of the three iconic figures of structuralism. Lacan, in particular, I found completely obscurantist. His heady mixture of surrealism, structural linguistics, and psychoanalysis seemed to me to be a form of linguistic idealism, one that implied that human subjectivity did not exist apart from language (Morris 2014A: 650-658).

Although I was particularly interested in Levi-Strauss's writings on animal symbolism, my Ph.D. thesis—an ethnographic study of the socio-economic life of a group of South Indian foragers, the Malai pantaram—hardly engaged with the theoretical debates that so fascinated anthropologists in the 1970s. My ethnographic study—which I situated almost as an after-thought within the tradition of cultural ecology (Steward 1955)—attempted to describe and explain the social life of the Malaipantaram in terms of two broad sets of historical factors. These were:

1) The imperatives of (or rather the possibilities offered by) their existence as a nomadic forest people and their food-gathering economy, and

2) The cultural setting with respect to their encapsulation within a pre-industrial state. I described at length both their social life within the forest environment—which combined an egalitarian ethos and an emphasis on respecting the autonomy of the individual—and their complex relationship with the wider Indian society. I therefore illustrated how the Malaipantaram endeavoured to retain their independence and cultural integrity despite being ridiculed and harassed by the caste communities of the plains, the exactions of the Indian state bent on their development and settlement, and the intrusions of a mercantile capitalist economy focussed around the trade of minor forest products. Long before James Scott's seminal study of south-east Asian tribal peoples (2009), I was thus indicating how the Malaipantaram were practising the "art of not being governed" (Morris 1982, 2014B: 217-237).

During the final decades of the last century, when I was engaged in ethnographic researches with respect to the matrilineal peoples of Malawi, structuralism came to be regarded as passé. In its stead anthropology suddenly became besieged (if that is the right term) by two very contrasting ideologies. These are perhaps best described under the rubrics of postmodernism and sociobiology.

I discuss these two extremist, or rather one-sided, currents of thought below, for the postmodernists wished to transform anthropology and ethnography into a literary enterprise, while the sociobiologists (or Neo-Darwinists), in contrast, envisaged anthropology, and the humanities more generally, simply as a branch of the biological sciences.

Anthropology and the Human Paradox

Scholars of diverse intellectual backgrounds have long reflected on the fact that there is an essential "paradox" or "contradiction" at the heart of human life. For there is an inherent duality in social existence: humans are intrinsically a *part* of nature, while at the same time, through our conscious experience, intense sociality, and complex symbolic culture, we are also in a sense *separate* from nature. The social ecologist Lewis Mumford wrote of humans living in "two worlds"—the natural world and what the Roman scholar Cicero referred to as our "Second Nature"—human social and symbolic life which is "within" first nature. Thus Mumford concluded that humans are in a sense unique, the only creatures that live "a twofold life, partly in the external world, partly in the symbolic world he has built up *within* it" (1951: 48; Cicero 1972: 185). Humans thus have a dual existence and what was important with regard to human life, Mumford contended, was to remain fully *alive* on both "the plane of organic existence and on the plane of symbolic participation" (1944: 147).

The phenomenologist Edmund Husserl was likewise always intrigued by the essential duality or ambiguity at the heart of human existence. As he put it, there is an intrinsic "paradox" in human life, namely "that of humans as world constituting subjectivity, and yet as incorporated in the world itself" (1970: 182). Thus humans, in a "natural-objective" sense are real living entities in the material world; but at the same time this is constituted or given meaning by human subjectivity.

No scholar expressed this existential "contradiction" or "paradox" at the heart of human life more fervently than the radical humanist Erich Fromm, who viewed it as the "essence" of human life. On the one hand, Fromm wrote, as natural beings, humans are intrinsically a part of nature, and as such are subject to physical laws and controlled by physiological imperatives. Humans as organisms have specific biological needs for food, water, shelter, and protection, as well as having sexual and psychological needs. On the other hand, through "self-awareness, reason, and imagination", as well as through symbolic culture, humans "stand apart" from nature. Indeed, Fromm writes that humans are something of an anomaly in the natural world—a "freak of nature" (Fromm 1955: 353). Like Marx, however, Fromm viewed these two aspects of human life not as implying a dualism, but rather as dialectically interrelated, and emphasized the need for humans to find ways of being "at home in the world", and of restoring a new sense of "unity" with the natural world and with other humans (1955: 25; Morris 2014A: 361-363).

It is important to recognise that even to think at all, as John MacMurray reflected (1957: 92), is to discriminate, and to make some kind of distinction between not only concrete entities in the world, but also between the different *aspects* of human life. In making such distinctions this does not necessarily imply what Jacques Derrida described as "logocentrism" (1976), the setting up of radical dualisms. Within Western philosophy, for example, *distinctions* are often made between existence and human consciousness, between the natural world (nature) and human social life (culture), between the objective world (reality) and subjective experiences (appearance, or phenomena), between the empirical (perceptual) and the transcendental (intellectual) realms of understanding, and, as Karl Marx famously denoted, between human life (action) and consciousness (reflective thought) (Marx and Engels 1965: 38).

I have always felt that these are valid and important distinctions (repeat, *distinctions*): that they have an axiomatic quality and are probably of universal significance, even though these distinctions may be expressed, in different societies, in extremely diverse ways. The tendency of Neo-Kantian scholars, positivists, literary theorists, and postmodern anthropologists to either repudiate these distinctions entirely, or to collapse them into some woolly abstraction such as "experience" or "network" seems to me rather unhelpful, if not just plain obfuscating. Even more contentious is to (*mis*)interpret these distinctions, such as that between the natural world and human culture (or between life and consciousness) as implying a radical dualism, that is, a Cartesian metaphysic.

What is needed is a theoretical approach that embraces both these *aspects* of human life in an integral or organic unity, with a *dialectical* sensibility. This implies neither reducing existence (nature) to consciousness (culture) as with the cultural idealists (or social constructivists)—which often verges on mysticisms; nor reducing consciousness (or culture) to existence (nature) as with the positivists and eliminative materialists. An alternative *dialectical* strategy has long been explored within the Western Intellectual tradition—from Marx to Merleau-Ponty—or at least among those scholars who have a sense of dialectics (Morris 2014A).

Anthropologists have long recognized the duality of human existence, that humans both inhabit, think about, and represent the world in which they find themselves. Anthropology was essentially a child of the eighteenth-century Enlightenment, a period that saw the flowering of both the empirical sciences and philosophical naturalism (see chapter three below). The key figures in its foundation were the following scholars: Jean-Jacques Rousseau, Charles-Louis de Montesquieu, Immanuel Kant, Adam Smith, Adam Ferguson, David Hume, and Johann Herder. Kant, in particular, is noteworthy. Although well-known as a philosopher, and indeed widely regarded as one of the greatest of Western Philosophers, Kant spent more than twenty years giving lectures on anthropology, as well as having a deep interest in physical geography. According to Herder, he was a lively and inspiring teacher. At the age of seventy-four, in 1795, Kant published a seminal work, *Anthropology from a Pragmatic Point of View*. By "pragmatic" what Kant intended was the use of anthropological knowledge to further human enlightenment and to widen the scope of human freedom, especially from religions, dogmas, and political oppression, and thus to advance the "dignity of human nature" (2007: 31). What is significant is that Kant claimed that the fundamental concern of philosophy was to address the question "what is a human being", or, in contemporary parlance "What does it mean to be human?" It was a question that should be addressed, Kant felt, not by speculative philosophy nor by engaging in "scholastics" but rather empirically, through observations of everyday social life, thereby developing a "science of the human being"—anthropology (2007: 227).

Kant long ago recognized that there were three distinct conceptions of the human subject—as a species-being, a generic subject, humanity (*mensch*); as a psychological (personal) being, with a unique individual self (*selbst*); and, finally, as a social being, a member of a particular group of people (*volk*). Kant's own focus however, was on humanity as a generic being, or as a universal category, and he described humanity in terms of several bio-psychological faculties or dispositions—for example, understanding (cognition), sensibility (sense perception), memory, desire, imagination, passions, affects, as well as the "faculty of using signs" (2007: 300). Like Aristotle, Kant defined the human species as an "earthly being endowed with reason" (2007: 23), thus emphasizing the duality of human existence.

As is well-known, Kant's own student and friend Johann Herder was highly critical of Kant's emphasis on universal reason and universal faculties. For Herder felt that this tended to downplay the importance and specificity of individual human cultures (*volk*) and of human language, as well as tending to ignore the poetic and emotional aspects of human life. But anthropology cannot be viewed simply as a "romantic rebellion" against the Enlightenment (Morris 1986).

Anthropology, as it came to develop over the years, has always embraced the duality of human existence, and has always expressed a "dual heritage". Both Karl Popper (1992: 69) and Mario Bunge (1998: 47) have described anthropology as the key social science. For despite its diversity—noted above—anthropology has a certain unity of purpose and vision. It is unique among the human sciences in both putting an emphasis and value on cultural difference (Herder), thus offering a cultural critique of industrial capitalism and much of Western Culture, *and* in emphasizing people's shared humanity (Kant), thus enlarging our sense of moral community, and placing humans squarely "within" the natural world. Anthropology has therefore always placed itself – as a comparative historical science – at the "interface" between the natural sciences (specifically biology) and the humanities. In many ways it is an *inter*-discipline, held together by also placing an important emphasis on ethnographic studies. Drawing therefore on both the Enlightenment and Romanticism, anthropology has always had a "dual heritage", combining both humanism and naturalism, interpretive understanding (hermeneutics) and scientific explanations of social and cultural phenomena. I shall discuss this dual approach to social life more fully below in Part Three.

Sadly, towards the end of the last century this "dual" heritage, which combined humanism (hermeneutics) and naturalism (science), came under attack from two types of extremists.

On the one hand, under the rubric of *postmodernism*, cultural or literary anthropologists came to completely repudiate the Enlightenment heritage of anthropology. They thus attempted to reduce anthropology (and ethnography) to literature (hermeneutics)—or even to autobiography (Marcus 1995). This reduction proceeded to include a rejection of history and social science; embraced dubious moral and cognitive relativism; and became increasingly obsessed with language, symbolism, and ritual.

On the other hand, some anthropologists went to the other extreme. Influenced by *sociobiology*, especially its offshoots memetics and evolutionary psychology, they advocated a form of reductive materialism. This involved an attempt to explain complex social phenomena, particularly religion, purely in terms of cognitive mechanisms, and to view the human subject as a "meme machine", merely a receptacle by means of which "selfish" genes and disembodied autonomous "memes" (cultural traits) replicated themselves. I shall offer a critique of sociobiology and its reductive materialism (naturalism) later in the manifesto (see chapter six), but it may perhaps be useful here to expand a little on the postmodernist phase in anthropology. For we are now being informed that no self-respecting anthropologist or social theorist today would welcome the label "postmodernist" (Hann 2010: 117). Nevertheless, the postmodernist ethos still has a pervasive influence in anthropology, particularly in relation to ethnographic studies.

Postmodern Anthropology

During the 1980s postmodernism became all the rage in anthropology. Scholars who only a decade earlier were making a fetish out of science or structural Marxism suddenly embraced postmodernism with an uncritical fervour. Nobody initially seemed to know what exactly postmodernism entailed, and one anthropologist admitted having guilty feelings about using the term (Fabian 1994: 103), but its impact on anthropology—whose essential insights, in fact, it largely appropriated—was quite remarkable. Viewed as a critical movement within anthropology—earlier anthropologists being depicted(or caricatured) as crude positivists or naïve Cartesian rationalists (or both!)—postmodernism was described as creating a "sense of disarray" within the discipline. A "crisis of representation" (or ethnographic realism) was declared to have arisen, and postmodernism, we were informed, had completely shattered the epistemological foundations of anthropology, even seriously questioning the legitimacy of studying other cultures (Clifford and Marcus 1986; Nencel and Pels 1991; Hastrup 1995).

As an intellectual ethos, postmodernism, it seems, was largely derived from the oracular writings of Friedrich Nietzsche, Martin Heidegger, and Ludwig Wittgenstein (all political reactionaries), as mediated via a number of postmodern philosophers and social theorists. Jacques Derrida, Richard Rorty, Jean-Françoise Lyotard, Michel Foucault, and Jean Baudrillard are among these scholars usually depicted as exemplars of this diffuse and rather inchoate intellectual movement or ideology, whatever their own thoughts about the matter (Sim 2005).

Noteworthy is the fact that Gilles Deleuze (and his colleague Felix Guattari) always fervently denied that he was a "postmodernist". For Deleuze was never attracted to hermeneutics (or phenomenology), nor obsessed with semiotics, and, as Alain Badiou suggested, one of the key motifs of Deleuze's materialist philosophy was to "get away from the obsession with language" (2009: 117; Morris 2014A: 693-723). Nevertheless, there are "postmodern" elements within Deleuze's writings, for he was, like Althusser, fervently opposed to history, dialectics, and the human subject. He was essentially a reductive materialist (see Deleuze and Guattari 1988).

There was much that was valid and important in the postmodern critique, even though postmodern anthropologists suffered from a kind of historical amnesia with regard to their own subject. The discipline had a long tradition of "anthropology at home" (Lewis 2014: 11), had always emphasized "difference" (the diversity of local cultures), sometimes to the point of absurd exaggeration, and in studying non-western cultures had always been inherently "reflexive" (Hastrup 1995: 49-50). Although exhorting realist anthropologists to adopt a reflexive or critical awareness of their own discipline, the postmodernists themselves hardly evinced much reflexivity with regard to their own uncritical embrace of literary (postmodern) theory.

Among the key themes highlighted by postmodern anthropologists the following are noteworthy: an emphasis on the historicity of being and cultural life; the critique of Cartesian metaphysics and the transcendental subject outside of society and nature; the undermining of the dualistic opposition between humans (culture) and the material world (nature); the importance of hermeneutic understanding and thus an opposition to crude positivism; the stress that there is no unmediated relationship between language (or consciousness) and the material world; the notion that social experience (the human life-world) forms the basis and background of any theoretical standpoint (that humans are both practical and contemplative beings); and, finally, the problematic nature of instrumental reason and the problem of equating truth with modern science (scientism) (Morris 2014B: 27). However, enchanted by literary theory and anxious to affirm their own importance and originality, postmodern anthropologists served to forget that all these issues had been articulated for more than a century by numerous scholars from a wide range of intellectual backgrounds: pragmatists, Marxists, Neo-Kantian scholars, and evolutionary biologists, as well as social and cultural anthropologists.

For many anthropologists, including myself, most of the ideas expounded by the postmodernists with such fervour were—to put it in prosaic language—simply "old hat". Indeed, one scholar has suggested that postmodernism was largely an exercise of putting "old wine into new bottles" (Lopston 2001: 25).

Many scholars have suggested that what the postmodernists and literary theorists tended to do was to adopt certain important ideas and insights derived from anthropology and sociology and to take them to extremes – the strategy of "moving to the limit", thus ending up with intellectual positions that Pierre Bourdieu aptly described as "simply absurd" (2004: 26). It involved as Andrew Sayer writes, the tendency of postmodernism scholars to "flip from naïve objectivism to relativism and idealism, from totalities to fragments, and from ethnocentrisms to new forms of self-contradictory cultural relativism" (2000: 79).

The nihilistic ethos of postmodernism thus came to the characterized by the following four tenets. Firstly, as (supposedly!) we have no knowledge of the world except through language or "descriptions" (to use Richard Rorty's [1982: 197] term) the "real" was conceived as an "effect" of discourses. Anti-realism was thus extolled, and we were informed that there was no "objective reality". As the symbolic anthropologist Mary Douglas wrote: "All reality is social reality" (1975: 5), or as Jacques Derrida famously put it: "there is nothing outside the text", arguing that humans lived fundamentally in a symbolic world: "one that denies us the reality we seek". Indeed, Derrida even suggests that nature—that which the term denotes—has "never existed" (Derrida 1976: 158-159; Mikics 2009: 2).

Derrida was later to plead that he had been misunderstood (so meaning is not, as he claimed, completely indeterminate!) and that he did not doubt the reality of the material world, nor wish to put the "whole world into a book". He was merely advocating a "textualist" approach, implying that there was nothing beyond the "context" (Rotzer 1995: 45).

Not surprisingly, Foucault described Derrida's prose style as a form of "obscurantist terrorism", in that Derrida's writings were so allusive and obscure that it was difficult to understand *exactly* what he meant. Derrida could then, as a savant, heap contempt on his critics for failing to understand him! (Lehman 1991: 77).

But the important point is that postmodernists tended to propound a neo-idealist metaphysics and thus to repudiate realism (Tyler 1986: 129): the notion that the natural world is an objective reality and has an existence independent of human consciousness (or language).

Secondly, the postmodern philosophers emphasized that there was no unmediated relationship between consciousness (or language) and the material world. To assume such a one-to-one mimetic correspondence, they dabbed objectivism or logocentrism as the "metaphysics of presence" (Derrida 1976: 49), or as viewing knowledge simply as a "mirror" of nature—Wittgenstein's early picture theory of language (Wittgenstein 1921; Rorty 1980).

That our perceptions and experiences of the world are always to some degree socially mediated had, of course, been recognized by anthropologists and sociologists ever since Marx's reflections on Feuerbach's materialism. But postmodernists again took this insight to extremes, and not only posited no relationship at all between language and the world, but came to espouse an epistemological (and moral) relativism. Truth was either repudiated entirely—for reality was only an "illusion" (Tyler 1986: 135)—or it was viewed simply as an "effect" of local cultural discourses (Rorty 1980; Flax 1990), or truth would be "disclosed" or "revealed" by elite scholars through poetic evocation (as with Heidegger 1994). Truth as correspondence with reality was thus repudiated, or even openly derided. As knowledge is always historically and socially situated, there can be, the post-modernists argued, no universal truths or values. Both objective knowledge and empirical science (including the social sciences) were thus repudiated, along with universal values.

Thirdly, postmodernism offered a welcome critique of the transcendental ego of Cartesian rationalism (and phenomenology)—the epistemic subject who stands outside both history (society) and nature—while a critique of the abstract individual of liberal political theory was also offered. What they failed to recognize was that both these conceptions of the human subject—transcendental and abstract—had been critiqued, indeed lampooned, by Karl Marx, social anarchists like Michael Bakunin and Peter Kropotkin in the nineteenth century (Morris 2014B: 179), and by anthropologists and sociologists. Indeed, as Derek Layder suggested, the inherently social nature of human life had been taken for granted as a fundamental premise of the social sciences, including anthropology, for the past hundred years (1994: 17).

But, emulating the structuralists, postmodernists again went to an extreme, announcing the "dissolution" of human subjectivity and the "end of man" (humanity), while declaring the whole idea of a self was a "fiction". Human agency was thus repudiated entirely, the human subject being viewed simply as an "effect" of ideology (Althusser), power (Foucault), or discourses (Derrida).

Finally, there was the wholesale rejection of the "metanarrative" of the Enlightenment (Lyotard 1984), which included, of course, such important cultural perspectives as evolutionary biology, radical forms of socialism, human rights discourses and all theories of emancipation, palaeontology, and universal human history. In its stead there was a strident celebration of the "postmodern condition", with its emphasis on fragmentation, cultural pastiche, alienation, nihilism, polyphony, and nomadism. This describes, however, not so much a new postmodern paradigm, or a model for a postmodern ethnography (as Tyler affirms), but rather the cultural effects or even the "logic" of global capitalism (Jameson 1998: 20).

Moreover, aping the style of Heidegger, many postmodern scholars expressed themselves in the most obscure and impenetrable jargon. An example from one of the doyens of postmodern anthropology will suffice: "No nomadic entities, each a complete bundle of differences succumbing in co-mingled self-loss, but correlative, correspondence incompletions having no fixed identity even in the character of their respective completion nor in the moment of their coupling co-pullal-ations...." (Tyler 1991: 79)— enough!

This is offered as a critique of the totalizing transcendental ego (perversely identified with realist anthropology!) and as a sample of poetic anthropology. It's hardly poetry? (For a critique of Tyler's postmodern anthropology, see Lett 1997: 12-16.)

Thus postmodernist scholars came to exclaim with some stridency the "dissolution" or the "erasure" or the "end" of truth, reason, history, nature (significantly, always put in inverted commas!), the self, society, science, and philosophy—misleadingly identifying all of these terms with conceptions that are transcendental, ahistorical, and absolutist. They thus appear to see nothing between the so-called "god's eye" point of view, a transcendental perspective beyond time and space, and local discourses, supposedly fragmented and indeterminable (Hollinger 1984: 81). In the process a sense of common humanity, of human capacities, of human praxis, and of human history is lost. There is no sense of a human life-world, "an infinite surrounding world of life" common to all people, as Husserl expressed it (1970: 139), that is prior and district from both cultural world-views and transcendentalism.

Yet in their rejection of history, in reducing social reality to discourses, in their epistemic and moral relativism, in their dissolution of human subjectivity or any sense that humans may have an *organic* unity, many scholars have remarked that there is an "unholy alliance" between postmodernism and the ethos of global capitalism, specifically the triumphalism of neoliberalism (Wood and Foster 1997; Kohan 2005).

It hardly needs mentioning that postmodernism has since its inception in the 1980s been the subject of numerous incisive critiques (see, for example, Gellner 1992; Bunge 1996; Callinicos 1997; Sokal and Bricmont 1999; Detmer 2003). On the undoubted merits of an earlier generation of (realist) anthropologists see Benthall 1995: 3; H. Lewis 2014: 1-26)

In this manifesto of dialectical or evolutionary naturalism, which aims to provide a basis or "grounding" for ethnographic realism, I shall affirm, contra to postmodernist rhetoric, the salience of a realist and materialist metaphysic, and the crucial importance of such conceptions as truth and representation, human agency, and empirical knowledge. I will therefore specifically outline and advocate anthropology as a historical and humanistic science, one that upholds and combines both humanism (hermeneutics), with its focus on lived experience, and naturalism (empirical science).

But first I shall devote some discussion to two topics that in many ways provide background material in the understanding of dialectical naturalism: Aristotle's down-to-earth philosophy and the eighteenth-century European Enlightenment.

Chapter Two

The Greek Legacy

With the human "paradox" in mind, this chapter focusses specifically on the two prominent philosophers of classical Greek culture: Plato and Aristotle. Essentially a mystic and religious thinker, Plato placed an emphasis on the spiritual dimension of human life. He tended therefore to downplay, or even denigrate, its material aspects. In contrast, Aristotle had a much more "down to earth" philosophy. Fundamentally a biological thinker, it was Aristotle who laid the foundations for what Lewis Mumford and Murray Bookchin described as an "organic worldview", a precursor of the dialectical naturalism advocated in this manifesto. But first, I will present a brief introduction to philosophy as a cultural worldview, or cosmology.

Philosophy as A Worldview

The term philosophy derives from the ancient Greeks and literally means the "love of wisdom" (*philia*: affection, love; *Sophia*: wisdom, knowledge). It is essentially a form of rational inquiry into the nature of human existence, one that aimed to provide both a comprehensive understanding of the nature of the world we inhabit and the place of humans within it. For the Greeks, philosophy therefore entailed both the acquisition of theoretical knowledge (*episteme*)—a systematic and critical enquiry into the nature of reality—and practical wisdom (*phronesis*), the use of such knowledge in order to live a good and ethical life. Knowledge of the world in all its forms, both material and spiritual (cultural), and ethical were thus, for the Greeks, not rigidly demarcated.

With the development of philosophy within Western culture, a clear distinction came to be drawn between "natural philosophy", which was identified with science and the acquisition of knowledge, and "moral philosophy", which covered other branches of philosophy, specifically politics and ethics. At the present time several branches of philosophy are generally recognized, including metaphysics, the study of the nature of reality (being, the universe) as a whole; epistemology, the study of the nature and scope of human knowledge; ethics, the study on those moral values that guide human action; logic, an enquiry into correct forms of reasoning or argument; and, finally, aesthetics, the analysis of values and judgements related to the arts. Ontology is usually described as a branch of metaphysics, one that focusses specifically on the nature of really existing things (from the Greek *onta*: thing).

Ontology has recently become a fashionable term among anthropologists, who have suddenly discovered that there is a material world of things beyond language and the "text". My focus in this manifesto will be on ontology, for all my ethnographic research over the past forty years has focussed on the relationship between people—mainly the matrilineal peoples of Malawi—and the natural world—specifically with respect to its biodiversity, its fauna and flora.

As philosophy offers humans a general frame of reference, a world outlook that gives meaning to human life and provides a guide to their actions, it is often considered synonymous with the concept of "worldview" (German: *weltanschauung*) (Lewis 1962: 15). Indeed, Karl Popper suggested that all philosophical problems were essentially focussed on the problem of cosmology. He characterized this problem as "the problem of understanding the world—including ourselves, and our knowledge, as part of the world. All science is cosmology...." (1959: preface).

What is significant about Western philosophy is that many of its iconic figures within this tradition have been mathematicians and/or fundamentally religious thinkers—usually both! For example, Plato, Descartes, Leibniz, Kant, Whitehead, Wittgenstein, and Husserl were all accomplished mathematicians, and embraced some form of religious metaphysic (spiritualism). These ranged from mystical panentheism to protestant Christianity. Small wonder that Alfred North Whitehead famously described Western philosophy as a series of "footnotes" to Plato (1929: 53). It is, however, important to recognize that not every Western philosopher was enchanted with mathematics—Aristotle, though interested in logic, wrote little about mathematics—nor were they necessarily engaged in the "quest for certainty", as Dewey (1929) well expressed it. Equally important, we need to recognize also that materialism, as a realist and coherent metaphysic, has long been an important part of Western philosophy. In fact, materialism had its origins long ago in the atomic theory of Leucippus, Democritus, and the Epicurean School of Philosophy that was given its classical expression in Lucretius' wonderful philosophical poem *On the Nature of Things* (*De Rerum Natura*), written in the first century BC. Although derided, marginalized, and often disguised, as in Baruch Spinoza's pantheistic treatise *Ethics*, philosophical materialism has always been a constituent part of Western philosophy (Roy 1940).

Although the Greeks seem to have made a distinction between philosophy and religion (*theologike*), philosophy in some form is certainly found in all human societies. For example, the *Analects* of Confucius, *Rig Veda* of Hinduism, the Buddhist scriptures, and Plato's *Timaeus*, though sacred texts, all offer philosophical reflections on the nature of the Universe and on the frailty of human existence, and suggest ethical guidelines in relation to social life. As with religion, few will deny that philosophy, as a system of ideas, is not also universal among humankind but also integral to human culture. But it is also well to remember—as is clear from Plato's dialogues and many sacred writings—that sceptical attitudes towards cultural beliefs and practices, whether religious or philosophical, have probably been expressed in all human societies, and that *naturalistic* conceptions of the world also have a long history. As Kay Milton has suggested: "There are often diverse ways of understanding the world even within a small village community." (1996: 67).

This is even more so when considering complex civilizations such as those of China, India, Islam, or the Aztecs. For example, although Indian culture is often misleadingly equated with Hinduism and viewed as giving expression to "universal spirituality", this portrayal of Indian culture is (as I have argued elsewhere) something of a "myth". For several of the recognized schools of Indian Philosophy—especially Samkhya and Nyaya—were essentially materialist philosophies, and express, like early Buddhism, an atheistic viewpoint (Morris 1994: 70-74). The cultural configurations of all human societies—particularly agrarian civilizations—are complex, multifaceted, and often express diverse philosophical perspectives.

But in all human societies, two modes of thought (or philosophy) are, I think, always to some extent in evidence. The first, empirical naturalism, is expressed in practical skills, economic activities. The second, religious cosmology, provides, in varying degrees, all-encompassing systems of meaning. These two aspects of social life were well described by the anthropologist A.R. Radcliffe-Brown long ago, when he affirmed that "in every human society there invariably exist two different and in a certain sense conflicting conceptions of nature. One of them, the *naturalistic*, is implicit everywhere in technology, and, in our twentieth-century European culture, has become explicit and dominant in our thoughts. The other, which might be called mythological, or *spiritualistic* conception, is implicit in myth and religion, and often becomes explicit in philosophy." (1952: 130).

An emphasis hardly needs to be made that all humans have, in varying degrees, their own unique philosophy, usually implicit, and to a large extent derived from the societies and social groups to which they belong. Indeed, it has been suggested that no person could function in the world without some kind of philosophy to serve them as a practical guide in their everyday activities (MacMurray 1933: 9; Peikoff 1991:2).

In his defence of a realist metaphysic, Mario Bunge has expressed this well. He suggests that every person has some cosmology, or other philosophical standpoint, usually tacit rather than explicit, in order to "navigate" in the world. Such a cosmology has both a conceptual function and a practical function. It provides a framework to make sense of the world about us, and to provide guidance in life in order to "formulate goals, choose means, design plans and to evaluate all (actions)" (Bunge 2001: 27-28).

But before a discussion on Aristotle's philosophy, I want to introduce here the most famous of all Western philosophies, that of the Greek mystic Plato, which continues to be pervasive in Western thought.

Plato's Spiritualist Metaphysics

In the Vatican there is a famous painting by Raphael entitled "The School of Athens". It depicts the older, grey-haired Plato pointing to the heavens, while the younger Aristotle—his student—beckons downwards towards the earth (Lewis 1962: 50). This painting captures, in essence, the paradox of the human condition.

In so doing, it also indicates the profound differences between the two Greek philosophers: Plato, a spiritualist or transcendental philosopher who seeks eternal truths, versus Aristotle, the empirical realist, who looks down at the natural world he inhabits and seeks truth in the earthly realm. Plato is fundamentally a religious mystic. Aristotle's philosophy, on the other hand, is deeply informed by his empirical, specifically biological, interests.

Plato, whose ideas were presented mainly in the form of dialogues, has had a profound impact on Western culture, particularly on philosophy and Christian theology. His metaphysics, on which I will focus here, was especially well expressed in his book *Timaeus*, written a year before he died in 348 BC. In this short text, Plato not only describes the creation of the world by a divine craftsman (*demiurge*) who creates order out of chaos, but also clearly outlines his own dualistic metaphysic. For Plato suggests that the world, or cosmos, consists of two completely separate "realms": the "realm of becoming", as is experienced in everyday life, and a spiritual or divine "realm of being". The visible and tangible "world of becoming" is one of flux and change, of disharmony, of disorder (chaos), and of perceptual experience, the concrete particulars that we apprehend through the senses. The "realm of being", in contrast, is a divine world of universal forms or ideas (*eidos*). These forms are perfect, eternal (timeless), and immutable, and can be known only through intelligence, specifically through detached, contemplative intuition (*theoria*).

One has to stress here—something lost of postmodern and hermeneutic scholars—that for Plato there is a clear distinction between observation (visual perception), and "apprehending (the forms) by intelligence with the aid of reasoning" (1965: 40). Significantly for Plato, the latter capacity of philosophical intuition was restricted to only elite aristocratic scholars. In a later period, as Neo-Platonic mysticism, it took the form of *"gnosis"*, the intuitive, esoteric knowledge of a spiritual or divine realm (or being).

It is important to recognize also that Plato was an objective idealist, and during the medieval period was considered a *realist*, because the spiritual forms (or universal ideas) he postulated have an ontological status. Thus the concepts and ideals have an existence, or reality, independent of the perceiving mind (Lewis 1962: 45). Given this spiritualist metaphysics there is thus some truth in the suggestion that Plato's philosophy was a rationalized form of shamanism (Dodds 1977: 209).

The relationship between the two "realms" of the cosmos—becoming and being—was expressed in the process of recollection, or "unforgetting" (*anamnesis*). In this process, ordinary material beings, such as an elephant, or cultural values, such as beauty or justice, were viewed as imperfect impressions or "copies" of pure, timeless forms which existed in the divine realm of being: the true reality (Grayling 2009: 236).

Several themes emerge from Plato's spiritualism and dualistic metaphysic. These themes may be briefly noted as follows: a high regard for mathematics; a complete contempt for sensual experience and empirical knowledge; faith in the human mind to attain perfect, absolute knowledge; and, finally, a critical disdain for the materialism of Empedocles and Democritus, which, in dispensing with the divinity, would, Plato felt, only lead to atheism and social disorder (Leroi 2014: 86). It is worth noting that Empedocles, although something of a poet and spirit healer, had suggested a form of materialism. He postulated that the world was composed of four basic elements: earth, water, air, and fire. These were experienced in the form of two contrasting sensations (wet/dry and hot/cold)—a thoroughly empirical theory. The four elements comprised a part of a symbolic schema that has deep resonance within Western culture (Morris 1994: 37-40; Rupp 2005).

Plato's conceptions of the human subject, which I have discussed more fully elsewhere, postulated a radical body/mind dualism, the immortality of the soul (*psyche*), and a veritable disdain for the human body. Thus, according to Plato, to gain pure knowledge (*theoria*), the soul (mind) must be freed (purified, released, separated) from the body—for the "soul of the philosopher very greatly despises the body, and flies from it, and seeks to be alone". Knowledge, and truth, for Plato, belong to a transcendent order, to the realm of eternal ideas or forms (*eidos*) (Plato 1954: 109-112; Morris 1994: 33-37).

The contention of one postmodern anthropologist, that realist anthropologists are adherents of Plato's philosophy of objective idealism with its dualistic metaphysics and its ultra-rationalism (Stoller 1989: 131-135), is completely fallacious. To equate the appearance/reality distinction, acknowledged by most ordinary people (!), with Plato's metaphysics is equally obfuscating. Realist anthropology, as affirmed by an earlier generation of scholars, adopted, of course, a metaphysics that was completely *opposed* to Plato's mystical idealism. Indeed, they drew from evolutionary naturalism, with its ratio-empiricist epistemology and its materialist ontology. Plato has more in common with the Songhay spirit-mediums (discussed by Stoller) than he does with most recent anthropologists. To what extent Saussure, and Levi-Strauss, can be described as Platonists, as Stoller following Derrida (1976, 1978) contends, is, of course debatable (see my own discussion of Levi-Strauss' structural anthropology 2014A: 615-639). But to equate the empirical science of realist anthropologists (such as Boas and Radcliffe-Brown) with Plato's ultra-rationalism is simply obfuscating.

We may now turn to Aristotle, Plato's most famous student, for it was Aristotle who, via his biological studies, essentially brought philosophy back down to earth.

Aristotle: Bringing Philosophy Down to Earth

Aristotle was described by his contemporaries as an intellectual jackdaw, for his interests were extremely wide-ranging. He was indeed an "intellectual omnivore". Aristotle has also been described as a universal genius. For he pursued knowledge—and the desire for knowledge he thought was a natural proclivity found in all humans—throughout his life. Aristotle's contributions to logic, political theory, ethics, the natural sciences, aesthetics, metaphysics, and the philosophy of mind were indeed astonishing: he laid the foundations for most of these areas of study. He had a profound influence on both European and Islamic thought, and until the rise of mechanistic science in the seventeenth century, Aristotle's writings had an almost hallowed status.

Although he has a reputation for being dogmatic and obscure, Aristotle in fact has an open and questioning style of philosophy. He always begins with simple questions or puzzles (*aporia*), and following the Socratic method he develops a critical argument in dialectical fashion. However, in contrast to Plato's lively dialogues, Aristotle's writings tend to be dry, concise, and prosaic. Bertrand Russell graphically (and misleadingly) described Aristotle's philosophy as "Plato diluted by common sense", remarking that "He is difficult because Plato and common sense do not mix easily." (1946: 184). But as many scholars have since emphasized, Aristotle was essentially a biological thinker: this Russell did not develop, and his Enlightenment critics—especially mechanistic philosophers like Frances Bacon—never really understood.

First and foremost a naturalist (in the nineteenth-century sense of the term), Aristotle expressed an "organic" way of thinking as Murray Bookchin (1995A) emphasized.

Aristotle, in fact, described himself as a *physikos*—"one who understands nature". Armand Leroi suggests there is a sense in which Aristotle's philosophy *is* biology, in that his ontology and epistemology were devised in order to understand the biological realm—specifically that of animal life (Leroi 2014: 8). Aristotle was essentially a philosopher of living nature.

To outline the basic tenets of Aristotle's philosophy I will focus on a number of key concepts that emerge from his writings: substance (*ousia*), structure or form (*eidos*), the "soul" (*psyche*), and his theory of causes (*aitia*).

Substance (ousia)

Like the pre-Socratic philosophers such as Thales and Anaximenes, whom Aristotle described as *physiologoi* (students of nature), Aristotle was primarily concerned with a systematic knowledge of things (*onta*)—natural phenomena. He was an advocate of what a recent scholar has termed an "object-oriented philosophy". Though Graham Harman (2010: 93-104) surprisingly proposed this as a new and innovative approach to philosophy, he completely fails to mention Aristotle!

Aristotle's philosophy always begins with the existence of individual things that belong to the material world and are perceived by the senses: a particular animal, or a particular person like Socrates. He thus only regards a particular thing as real in the full sense—as substance (*ousia*)—and substance for Aristotle primarily means concrete individuals. Substance, Aristotle writes in his *Metaphysics* is "thought to belong most obviously to bodies, and so we say that both animals and plants and their parts are substances; and so are natural bodies such as fire and water and earth and everything of that sort" (Ackrill 1987: 285). But essentially when Aristotle spoke of a substance he meant one particular kind of existent thing, a primary substance being an individual organism, while natural species are described as secondary substances (Ackrill 1987: 7).

Substance, Aristotle writes, is one of ten categories (*Kategoria*, meaning predicate) that can be applied to a subject. The others are quality, quantity, relation, location, time, position, state, action, and affection. These ten categories are referred to as "categories of being". But it is clear that for Aristotle the primary substance, the individual organism, was not itself a predicate, and that the various predicates (categories), though real enough, have no independent existence apart from primary substances. As Aristotle wrote: "Substance is primary in every sense—in formula, in order of knowledge in time. For of the other categories none can exist independently but only substance." (Ackrill 1987: 285).

It is, however, worth noting that for Aristotle, substance (*ousia*) came to have a wide range of meanings. It thus referred not only to individual organisms and natural species, but also to a subject, and to the underlying substratum (*hypokeimenon*) of a thing—"that from which a thing comes to be"—as well as to what it is, in terms of its essence or form (*eidos*) (Lloyd 1968: 129).

Aristotle always affirmed two basic assumptions. First, he held that natural things are always subject to change. Second, he held that *what* a thing "is" always has several meanings, as "things are said to be (to exist) in many ways" (Ackrill 1987: 82-89). Aristotle's philosophy was certainly empiricist, but it was also in many ways materialist, although he never doubted the existence and agency of a supreme spirit and deity (*theos*).

Aristotle has long been recognized as an advocate of hylomorphism, the theory that explains the genesis of an individual by starting from the union of its matter (*hyle*), and its form (*morphe*). This has been implied to mean the external "imposition" of abstract forms on an inert matter, as viewing natural species as fixed and immutable, and as explaining the genesis of the individual without reference to its relations or to its milieu. But as we note below, Aristotle did not in fact consider the material world to be "inert", and as a biologist recognized the importance of relations, which he considered one of the categories of being (on hylomorphism see Simondon 2012: 212).

Form (Eidos)

As we discussed above, Plato regarded the existence of ideas, or forms, as independent of particular things and as having an ontological reality of their own. Aristotle kept Plato's distinction between form and matter, but gave it an entirely different meaning. He suggested that any substance, any individual material thing, is in a sense composite. It consists essentially of two parts: matter (*hyle*, also meaning timber)—what it is made of—and structure or form (*eidos*). A table, for example, is made of wood, and it also has a particular shape or form that is intimately related, in Aristotle's way of looking at things, to its purpose or function. For Aristotle, the form or the essence of a particular thing is always *within* the material object. This is why Aristotle, with respect to universals (such as the colour red), has been described as an immanent realist (S. Mumford 2012: 23).

It has been suggested that Greek thought was characterized by two contrasting philosophical tendencies. One was materialist in temperament: it posed the question "what is it?" in terms of matter. In the other teleological tendency, questions were posed in terms of the purpose or function of a thing. As Guthrie wrote: the "primary opposition which presented itself to the Greek mind was that between matter and form, always with the notion of function included in that of form" (1950: 21). Aristotle combined both types of explanation, and his concept of form was strongly cast in a biological idiom, as teleology.

Aristotle wrote that there were two sorts of things called nature (physis): form and matter. As a student of nature (as he described himself), Aristotle was keen to embrace both. He was therefore critical of his tutor Plato for completely ignoring the empirical realm of becoming, and for focussing entirely on a transcendental, spiritual realm of eternal forms. Aristotle thus repudiated Plato's dualistic metaphysic. He was equally critical of the early materialist thinkers like Empedocles and Democritus, who focussed entirely on the material world, and thus, he wrote: "touched only very superficially on form" and on what something might become (Ackrill 1987: 97).

It is important to recognize that Aristotle, like other Greek philosophers, saw the world as a process, and that nature (*physis*) was conceived essentially as a source or cause of change. For Aristotle, both order and motion were intrinsic to the world, and the term used to describe the totality of things, the Universe as a whole, was *kosmos*. The term is derived from the verb which means "to order", or "to arrange". In some contexts, *physis* and *kosmos* refer to much the same thing, to nature as a totality; but in another, and more important sense, nature (*physis*), for Aristotle, seems to denote something that is *within* each natural object. Closely linked with the verb "to grow" (*phylein, phyton*), Aristotle affirms that in one sense nature (*physics*) refers "to the coming to be of things that grow" (Ackrill 1987: 270).

In "*Physis*" Aristotle considered both form and matter as modes of understanding things in the world. He concluded that the nature of something is "not the thing it is growing out of (matter), but the thing it is growing into. So the form is nature." (Ackrill 1987: 95). Thus Aristotle came to see things in the world as developing organisms, organisms in which matter was the potentiality (*dynamics*, potency), and its form (*eidos*) was the actuality (*entelekheia*), the mode of being of the thing. He labelled the process through which the form is realized with the term *energeia* (literally, putting to use). For Aristotle, all change is therefore seen as some kind of growth or development towards some purpose or end. This brings us to Aristotle's concept of the soul (*psyche*), which is essentially viewed as the form (*eidos*) of a living body (*hyle*).

Soul (Psyche)

Aristotle devoted a large part of his scientific work to the study of natural phenomena, particularly living things. His writings on the soul (*psyche*), as expressed in the classic treatise known by its Latin title *De Anima*, therefore must be understood as coming from a fundamentally biological thinker. To interpret this work, and the psychology expressed therein, in Platonic fashion, or to try to fit Aristotle's ideas into current debates on the philosophy of the mind, is to seriously distort the tenor of the work.

It has to be said at once that the translation of the Greek word "*psyche*" as "soul" is seriously misleading, as this term, for Aristotle, has a much wider meaning. Essentially it means the "principle of life": that which animates a living thing, whether plant or animal. It is thus a much wider concept than the terms "mind" and "consciousness", and completely lacks the spiritual connotations of the Hindu or Christian concept of the "soul". Indeed, what is striking about the study *De Anima* is that Aristotle is not really interested in the immortality of the soul, and is highly critical of the Pythagoreans and their theories of the transmigration of souls. Aristotle repudiates all dualistic theories of the soul, and argues that the body (*hyle*), and the soul (*psyche*) are dialectically interrelated. As he writes: "it is by their partnership that the body acts and the soul is affected, that the body comes to be moved and the soul produces motion" (1986: 142).

Aristotle thus comes to suggest two things: an intimate relationship between the body and the soul, and a tripartite conception of the soul itself in terms of three different faculties or powers (*dynameis*). The relationship between body and mind follows the distinction he makes between matter and form, potentiality and actuality. Soul is therefore seen as the "form" of the body. Aristotle thus writes: "It must be the case that soul is substance as the *form* of a natural body, which potentially has life, and since this substance is actuality, soul will be the actuality of such a body." (1986: 157; Ackrill 1987: 165).

Aristotle describes the soul as comprised of three faculties (or functions). He writes: "Now the soul comprises cognition, perception and belief states. It also comprises appetite, wishing and desire—states in general. It is the source of locomotion for animals, as also of growth, flourishing and decay." (1986: 153).

Aristotle speaks of the nutritive, perceptive, and imaginative (or intellectual) faculties. All living things have the intuitive faculty, and go on living as long as they are able to take nourishment. They have the capacity for reproduction, growth, and decay. Animals, in addition, also have the capacity for perception and spatial movement—this also implies feelings of pleasure, pain, and desire. Therefore, Aristotle suggests, the soul is neither without body, nor a kind of body. Instead it belongs to a body: it is the cause and principle of living potential. All living things have a soul—plants, fungi, animals, and humans—and when a living thing dies, its soul ceases to exist. Aristotle recognizes that all living things have these powers, or capacities, to varying degrees. But only humans have the faculty of thinking, understanding, and imagination.

Aristotle rejected Plato's dualism, and he appears not to have believed in personal immortality. He would probably have repudiated the Cartesian mind/body dualism. But he was also not a reductive materialist, for he was critical of the theories of Democritus and Empedocles. Aristotle was essentially a functional biologist. He has affinities with the American pragmatist John Dewey and the functional school of psychology (Leahey 1987: 49; for general accounts of Aristotle's theory of the soul see Lloyd 1968: 181-201; Ackrill 1981: 55-78).

What has to be recognized, of course, is that there is a deep contradiction between Aristotle's philosophical biology and his aristocratic politics. For in his book on *Politics* there is a sterling defence of the city-state and the institution of slavery: therein Aristotle also links the soul (and reason) to the aristocratic male who is alleged to have a natural "mastery" over the body, slaves, women, and animal life (Aristotle 1962: 68; on Aristotle's political psychology see Okin 1978: 15-98; Morris 1994: 47-48).

Causes (Aitia)

Aristotle never expressed any disdain for the everyday knowledge (*doxa*) of ordinary people. In fact, he always began his scientific investigations with a critical examination of popular wisdom and the current approaches to a particular subject matter. As he wrote with respect to understanding the nature of the soul (*psyche*): "It is necessary that we collate the opinions of as many of our predecessors as have given a view about the soul with the aim of adopting all sensible proposals." (1986: 132; Leroi 2014: 45).

But for Aristotle, an empirical or scientific realist, to attain any systematic knowledge (episteme) of any material thing we must go beyond everyday perceptual experiences and seek its underlying causes. As he wrote: "Wisdom is knowledge about certain causes and principles." (Ackrill 1987: 257).

But such knowledge, Aristotle argues, is not for any utilitarian end, for it is according to *wonder* that humans first began to philosophize (Ackrill 1987: 258). Aristotle calls this "free science", and suggests that there are four basic causes or kinds of explanation. These are, as he succinctly put it: "the matter, the form, the thing which effects the change, and what the thing is for" (Ackrill 1987: 105). Significantly, he noted that the last three coincide.

The four causes are thus as follows: the *material* cause, an account of the matter (*hyle*)—the stuff—of which living things are made; the *formal* cause, the form (*eidos*) which gives "an account of what the being would be"; the *efficient* cause—the primary source of change or movement; and finally, the *final* cause (*telos*), the understanding of things in terms of their purpose or function. When Aristotle emphasized their biological origins, Leroi suggests that each of these four kinds of explanations has its counterpart in the various branches of contemporary biology: material cause (physiology, biochemistry), formal cause (genetics *eidos* being considered akin to a "genetic programme"), efficient cause (developmental biology) and, finally, the final cause (evolutionary and functional biology—particularly with regard to the emphasis on adaptation) (Leroi 2014: 91-92).

To conclude this discussion of Aristotle's philosophy, and given my own ethnographic interests, we may offer some reflections on Aristotle's natural history. The history of biological taxonomy, wrote Ernst Mayr (1982: 49), starts with Aristotle, who is often depicted as an advocate of a subject-predicate logic, or the "logic of identity" that tends to conceive of an existent thing as requiring "nothing but itself in order to exist"—and thus to deny the crucial importance of relationships (Whitehead 1929: 64).

However, as we have emphasized above, Aristotle, unlike the mathematician Whitehead, was fundamentally a naturalist whose years in the eastern Aegean (347-335 BC) were largely devoted to natural history studies, particularly the two years he spent on the island of Lesbos (345-343 BC) (see Leroi 2014 for a fascinating account of the island and Aristotle's biology).

Although Aristotle expressed a natural teleology and conceived of the natural world as essentially eternal—with each "species" being essentially fixed and unchanging, possessing an inherent nature (*physis*) and form (*eidos*)—his approach to animal life was essentially empirical and ecological. In his well-known study *Historia Animalicum* (1965-1970) Aristotle was not primarily concerned with an exhaustive classification of animals, nor working through some deductive logic, but largely with recording the ecology, activities, dispositions, and morphology of animals. Aristotle thus never conceived of animals simply in morphological terms, or as self-existent isolated entities (does anyone!), but essentially has an ecological or relational perspective. He emphasized the relationships of various kinds (*genos*) of animals to the world. He noted and described, for example, the diverse feeding habits of various forms (*eidos*) of animals; the fact that they have agency and provide themselves with specific habitations; their varying dispositions towards humans; that cranes, ants, wasps, and bees are, like humans, essentially social animals; and that the morphologies of birds and animals clearly have functional correlates, in that birds of prey have talons and hooked beaks (Aristotle 1965-1970).

The main life-forms (*genos*) of animals recognized by Aristotle were the following: quadrupeds (*tetrapoda*), birds (*ornithes*), snakes (*opheis*), fish (*ikthyes*), insects (*entoma*) (the presence of blood being a key distinguishing feature of these groups), and various forms of marine life. Aristotle seems to have been very knowledgeable with respect to marine life in the lagoons on the island of Lesbos, being particularly fascinated with the biology of dolphins, and recorded over forty different species of fish (Morris 2004A: 15; Leroi 2014: 386-395).

First and foremost a naturalist, Aristotle always expressed a positive attitude towards the natural world, particularly towards animal life, for as he wrote: "In all natural things there is something wonderful." He continues: "We should approach the investigation of every kind of animal without being ashamed, since in each one of them there is something natural and something beautiful. And the end for the sake of which a thing has been constructed or has come to be belongs to what is beautiful." (Parts of Animals 1/5: 20; Ackrill 1987: 227).

Clearly Aristotle did not share that disdain for nature that David Abram (1996: 94) sees as intrinsic to European civilization. For Aristotle, however, the natural world was eternal and unchanging. Everything in nature had its purpose, and there was an essential unity between an organism's form (*eidos*) and its purpose or its function (*telos*). Aristotle was thus not an evolutionary thinker. But he was not a mechanistic philosopher either, and Francis Bacon's harsh criticisms of Aristotle stemmed from the fact that Bacon considered formal, final causes to have no legitimate place in scientific research—that natural-mechanistic philosophy should instead be concerned solely with the properties and movements of matter (i.e., with material and efficient causes). But as both Mayr and Leroi have suggested, the emergence of molecular biology has made Aristotle's principle of form (*eidos*) "respectable again" (Mayr 1988: 56-57; Leroi 2014: 359-372; on Aristotle's biology see also Gotthelf and Lennox 1987).

There is some truth in the contention that Aristotle not only firmly established the science of biology, but also the empirical sciences generally, and that in recognizing the different levels of biological organization Aristotle may also have been a materialist of sorts. He certainly advocated an empiricist approach to knowledge (Leroi 2014: 162).

Equally significant, the level of biological understanding went steadily downhill after Aristotle (Mayr 1982: 153), that is, until the emergence of evolutionary biology during the eighteenth-century Enlightenment. This was a period when a materialist metaphysic was firmly re-established, and so it is to the Enlightenment that I shall now turn.

But it is worth recalling that Aristotle implicitly expressed a number of key ideas that were adopted by later scholarship. Namely, he expressed a materialist ontology; a form of perspectivism, in recognizing that any particular thing could be viewed as existing in a variety of different ways; a distinction between the form of function (actuality) of a thing and its underlying substratum (on processes)—its potentiality, or what Deleuze was to describe as a virtual realm; a relational epistemology which emphasized that what a thing is is intrinsically related to other entities; and finally a critical or scientific realism that acknowledged the need to go beyond perceptual knowledge (phenomenology), and to seek the causes—an explanation—of phenomena.

What Aristotle lacked, of course, was a historical or evolutionary approach to the natural world. This was the missing link that the Enlightenment provided with the development of a historical or emergent materialism at the close of the eighteenth century.

It is worth noting, of course, that within the Greek legacy, we also need to acknowledge the materialist philosophy of Democritus and Epicurus. The importance of both scholars will be broached in chapter seven, in discussing Marx's dialectical materialism.

A last word. It has often been said that all Western philosophers were followers either of Plato or of Aristotle: one an advocate of religious mysticism (*theos, theoria*), the other empirical naturalism (*physis, episteme*). Needless to say, I have long been a fan of Aristotle, even though I abhor his statist and reactionary politics, especially his defence of slavery, gender inequality, and private property.

Chapter Three

In Defence of the Enlightenment

This chapter offers an account and a defence of the eighteenth-century Enlightenment. It consists of three parts.

In the first part I describe the legacy of the Enlightenment and focus on four key concepts that constitute its radical import, namely, the concept of reason, the politics of liberty, and the ideologies of individualism and universalism (humanism).

In the second part I discuss the emergence, in embryonic form, of a new metaphysic during the late eighteenth century, that of historical naturalism. This entailed an awareness of "deep time" with respect to the earth's history, and the development of an evolutionary theory. I suggest that a distinction needs to be made between the Enlightenment, as an intellectual tradition, and "modernity", whether the latter is viewed as a historical period or as a cultural perspective.

In the final part I discuss the "counter-Enlightenment" and respond to the various critiques of the Enlightenment, both in the past, by such conservative reactionaries as de Maistre and Gobineau, and in the present. I stress that many of the contemporary postmodernist critiques of the Enlightenment tradition are often misplaced, or are such crude misrepresentations that they give us a travesty of Enlightenment thought. The chapter as a whole aims to emphasize the continuing relevance of the Enlightenment in formulating a philosophy of dialectical naturalism.

The Legacy of the Enlightenment

It is now a common pastime among academic scholars—particularly literary theorists and postmodernist or "post-human" anthropologists—to express a blanket dismissal of the Enlightenment tradition. Indeed, it is highly fashionable to deride and denigrate the Enlightenment, even holding it (along with the faculty of reason!) responsible for the political horrors of the twentieth century. I shall return to this "dialectic" of the Enlightenment—as Adorno and Horkheimer (1973) famously described it—below; it will suffice to note here that postmodernist accounts of the Enlightenment tend to verge on "caricature" (Sim 2005: 207).

What then was the Enlightenment tradition?

The term Enlightenment—in German *aufklarung*—has two essential meanings. On the one hand it refers to the period of European history, roughly the eighteenth century, the "Age of Enlightenment" (*Siècle des Lumières*), as it was often described. On the other hand, it refers to a group of radical French intellectuals known as the "*philosophes*" who were focussed around the publication of a vast compendium of knowledge known as the *Encyclopédie*. Edited by Dennis Diderot, the *Encyclopédie* was not simply a dictionary of the arts and sciences, but, according to Diderot, it aimed to bring about a fundamental change in the human mode of thinking and to make people happier and more virtuous (Cassirer 1951: 14; Kramnick 1995: 17-21). Scholars specifically associated with the *Encyclopédie* include not only Diderot, but also Voltaire, D'Alembert, Montesquieu, Quesnay, and Rousseau. Closely identified with the French Enlightenment were also such luminaries as Helvetius, D'Holbach, Condillac, and La Mettrie. In fact, Julian de la Mettrie's *L'Homme Machine* (1746) and Paul D'Holbach's *Système de la nature* (1770) laid the foundations for what Marx and Engels were later to describe as mechanical materialism.

But the Enlightenment was not restricted to France, and scholars have written also on the British and American Enlightenments. For the essential ideas of the Enlightenment as a radical intellectual tradition were widespread. In Britain, Hume, Adam Smith, Godwin, Wollstonecraft, Ferguson, and Thomas Paine were also identified with the Enlightenment, and in the United States Benjamin Franklin, Elihu Palmer, and Thomas Jefferson were the key figures. In Germany, the philosopher Immanuel Kant famously wrote a short essay *What is Enlightenment* (1784), he declared that the motto of the Enlightenment was *sapere aude*—"have the courage to use your own reason" (Kramnick 1995:1).

The Enlightenment was, of course, a highly varied phenomenon, and as Ernst Cassirer noted, it is often treated as an "eclectic mixture" of very diverse elements, for in terms of basic metaphysics, politics, and religious affiliation, there were wide disagreements among Enlightenment thinkers. Nevertheless, there is a sense in which the Enlightenment had a certain unity, and scholars have described the Enlightenment as an ideology, or worldview, while its postmodernist critics repeatedly refer to the Enlightenment as a "project". In his defence of the Enlightenment, Tzveran Todorov (2009) writes of the "spirit" of the Enlightenment.

It is beyond the scope of the present manifesto to examine the Enlightenment in terms of its social and historical context—that of the eighteenth century. The focus instead will be on the Enlightenment as an intellectual tradition, one that continues to have a deep resonance among contemporary scholars (e.g., Bunge 1999: 129-143; Bronner 2004; Pagden 2013. For useful general studies of the Enlightenment see Hampson 1968; Munck 2000; Porter 2001; Himmelfarb 2008).

As a radical ideology, the Enlightenment was focussed around a number of key ideas or concepts. Those include reason and liberty, individualism and universalism, and a form of thought that we may describe as historical naturalism. I will discuss each of these concepts in turn.

The Concept of Reason

The central and unifying concept of the Enlightenment was that of reason, a natural faculty shared by all humans. This faculty emphasized the kind of thinking employed in science and philosophy and in the formation of empirical knowledge—what the Enlightenment scholars referred to as "common-sense" understanding. Reason is notably expressed in the natural curiosity that humans have for understanding the objective world, and in the discovery, through causal analysis, of the immanent connections between phenomena. It implied what a later generation of anthropologists would fervently extol as a "relational epistemology". The Enlightenment concept of reason was therefore empirical, not a priori or "pure", for, as Ernst Cassirer wrote: "The power of reason does not consist in enabling us to transcend the empirical world but rather in teaching us to feel at home in it." (1951: 13).

It is therefore important to make a clear distinction between the Enlightenment concept of reason and philosophical rationalism. Indeed, to describe the eighteenth century as the "age of reason" is something of a misnomer, for Enlightenment scholars consistently argued that experience and experiment, not a priori reason, were the keys to human knowledge (Porter 2001: 2). As Diderot succinctly described the acquisition of empirical knowledge: "Observation collects the data, thought combines them, and experiment verifies the result." (Hyland 2003: 112). Human knowledge, specifically scientific understanding, is thus the result of combining reason with empirical studies.

When the Christian theologian Alister McGrath writes that the "basic presupposition of Enlightenment rationalism is that human reason is perfectly capable of telling us everything we need to know about the world, ourselves and God (if there is one)" (2001: 181), he seriously misjudges the Enlightenment concept of reason and the philosophes' conception of human knowledge. Enlightenment thinkers were highly critical of the pure rationalism expressed by Descartes, Spinoza, and Leibniz, seeking to extend the boundaries of knowledge and to make knowledge more empirical, more concrete, more elastic, and more vital (Cassirer 1951: 7). Knowledge was thus seen as having two basic sources: the intellect and experience. Kant's famous study *Critique of Pure Reason* (1781) was essentially concerned with the integration of the rationalism of Descartes, Spinoza, and the German scholar Christian Wolf with the empiricism of Locke and Hume. Although he looked upon himself as a kind of Copernican figure who had created a new synthetic science, Kant was nothing of the sort: he remained essentially a religious thinker. His transcendental idealism implied a dualistic metaphysic, with the bifurcation of reality into a phenomenal and an (unknowable) noumenal realm. Thus the radical dichotomies between freedom and necessity, subject and object, consciousness and mechanism, and morality and science became ever more deeply entrenched in Kant's philosophy rather than being dialectically resolved and overcome (A. Wood 1984; Morris 1991: 45-57).

But crucially, Enlightenment philosophes were concerned to promote not only the scientific study of nature—the natural world—but also of human social life, and to foster the development of what may be described as the "critical spirit". This led to their questioning of all forms of knowledge based on religious revelation, mystical intuition, or recourse to the authority of traditional beliefs, and to stress the importance of free enquiry. People were encouraged to think for themselves, to be critical, and thus not to base knowledge on any form of authority

(Todorov 2009: 5). Inevitably, given that "reason" was the unifying and key concept of the Enlightenment, its significance was often exaggerated. During the French Revolution it was, rather pathetically, even given divine status as a "goddess" (Kramnick 1995: 168; Bunge 1999: 131).

A stress on the importance of reason did not, however, imply for the philosopher a devaluation or a denial of the importance of the imagination, the emotions, or the body. Scholars such as Diderot, Hume, and Adam Smith all emphasized the important role that sentiments, emotions, and passions played in human life. For the Enlightenment thinkers, human beings were not Cartesian disembodied rational egos, but rather both mind and body, rational and emotional, sensual and contemplative, practical and imaginative. They recognized long before the phenomenologists that humans were "embodied"! The important point was that the exponents of the Enlightenment were not, contra McGrath, pure rationalists, they simply emphasized reason rather than basing knowledge on moral judgements, subjective feelings, religious faith, mystical intuition, or cultural authority. Indeed, Peter Gay suggests that the Enlightenment was a "revolt" against rationalism (1966: 141; Yolton 1991: 445-46).

The Politics of Liberty

The second important concept of the Enlightenment was that of *liberty*, or what Todorov (2009) describes as the "principle of autonomy". For the Enlightenment thinkers placed a crucial emphasis upon the freedom of the individual. This meant, of course, that they were highly critical of aristocratic rule, and particularly the divine right of Kings, as well as the authority of the Catholic Church, which they adjudged to have a corrupting influence on people's lives.

But the philosophes were not anti-religion, for although D'Holbach and Diderot were atheists, many of the philosophes, including Voltaire, Paine, Rousseau, and Jefferson, were deists. But importantly, Enlightenment thinkers emphasized that religion should be a matter of individual reflection and conscience. They advocated both a spirit of religious toleration and secularism—the separation of state politics (power) from religious dogma. They therefore opposed all forms of theocracy (Gay 1966: 358-419; Outram 1995: 31-46; Todorov 2009: 57-61).

Enlightenment thinkers such as Diderot, Montesquieu, and Rousseau also condemned slavery, considering it to contradict the principle of human rights. One of the philosophes, Claude Helvetius, expresses this critique of slavery rather graphically when he wrote that: "Not a barrel of sugar arrives in Europe that is not tainted with human blood" (quoted in Hampson 1968: 10). Yet although the Enlightenment thinkers emphasized the values of humanism and universalism, it has to be recognized that some of the philosophes were not free of the racial prejudices of their time. For scholars such as Hume, Voltaire, and Kant all stressed the intellectual and cultural superiority of "white" Europeans, and viewed Western culture as the apotheosis of human civilization (Hume 1985: 208; Patterson 1997; Llobera 2003: 33-34).

The "politics of liberty" that the philosophes advocated stressed the importance therefore of the natural rights of the individual, and particularly the freedom of the individual vis-a-vis state power. This conception of politics was well expressed in the *American Declaration of Independence* from British colonial rule (1776) and in the *American Constitution*. This document affirmed that "all men are created equal" and that they are endowed by the creator with "certain unalienable rights", among which are "life, liberty and the pursuit of happiness". The function of government was to secure and protect these rights. It held that such state power should be derived from "the consent of the governed"—the people (Kramnick 1995: 449).

The *Declaration of the Rights of Man and the Citizen* by the French National Assembly (1789) reaffirmed this conception of politics, as did the rallying cry of the French Revolution—"Liberty, Equality, Fraternity". Significantly, the French declaration included the right of "resistance to oppression," something lost in the *Universal Declaration of Human Rights* (1948), which outlaws any attacks on the juridico-political order and allows states to declare a state of emergency, and thus to completely abrogate any universal rights (Bobbio 1996: 75-93).

But this emphasis on "liberty" and on the natural rights of men, given the historical context, was essentially restricted to "men of property" (the bourgeoisie). It thus did not extend to the labouring classes, women, African slaves, and Native Americans. Indeed, scholars like Voltaire and Diderot clearly felt that their Enlightened politics could best be advanced by some Enlightened monarch, such as Catherine II of Russia, and Frederick II of Prussia—both absolutist monarchs!

Nevertheless, the ideals expressed by the Enlightenment, as the "politics of liberty" came to be an important factor in movements for gender equality and universal suffrage, in the anti-slavery movement, as well as in struggles against colonial oppression. For such movements were inspired by the principles of the Enlightenment, particularly the emphasis on the equality of humans and the importance of human liberty (Bronner 2004: 66; Todorov 2009: 31).

Individualism

Given the emphasis on natural rights and the freedom of the individual, it followed that the philosophes embraced, to an important degree, the doctrine of *individualism*.

Acknowledging that the people of all cultures and races had the "same nature" (as D'Holbach expressed it)—or as David Hume wrote, "mankind are much the same, in all times and places" (Gay 1969: 169)—thereby emphasizing the psychic unity of humankind, the philosophes nevertheless tended to privilege and highlight the individual as an autonomous agent vis-a-vis the social context. The Enlightenment thinkers thus expressed a rather atomistic conception of human life and viewed the human being as a rather solitary subject, or, as Marx put it, they had a rather "abstract" conception of the human subject. Enlightenment thinkers—and Marx had in mind Adam Smith, Ricardo, and Bentham—essentially viewed the individual in an abstract sense, as the product of nature rather than human history, and thus downplayed the social aspects of the individual. For Marx, humans were intrinsically social beings, and could "develop into an individual only in society" (McLellan 2000: 381; Morris 2014A: 28-31). The human being was not simply an "isolated" individual, an exemplar of some "abstract humanity", but rather a historical and cultural being, enacting many social roles, or, in contemporary parlance, the foci of multiple identities.

With regard to Enlightenment scholars such as Helvetius and Bentham, an atomistic individualism was certainly evident. Claude Helvetius expressed a rather narrow biological and behaviourist approach to the human person. He stressed the primacy of sense perceptions and reduced all human motivations to the pleasure-pain principle. His book *De L'Esprit* (on the mind) (1758), though prescribed by the French government, was a rather crass and unoriginal expression of mechanistic materialism and psychological egoism.

This kind of utilitarian individualism was also expressed by the British philosopher Jeremy Bentham, one of the founders of the utilitarian school of philosophy, and of methodological individualism—the latter implying the rejection of the notion that social phenomena have emergent properties. Bentham described the human community as a "fictitious" body, and his theory has been described as being like that of Hobbes, one of "ethical egoism" (Lukes 1973: 100; Kramnick 1995: 306-314).

Yet there were inherent tensions within the Enlightenment, for other scholars strongly emphasized the social nature of the human subject and were critical of the Cartesian notion of the individual as a disembodied rational ego and of the Hobbesian materialist conception of the human person as a possessive and utilitarian individual. Both Montesquieu and Ferguson stressed the fact that humans were intrinsically social beings. As Ferguson put it, quoting Montesquieu: "Man is born in society and there he remains." (1995: 21).

Indeed, Adam Ferguson's *An Essay on the History of Civil Society* (1767), along with Montesquieu's earlier *The Spirit of Laws* (1748), were of central importance in establishing sociology and political theory as social sciences. Adam Smith's *An Enquiry into the Nature and Causes of the Wealth of Nations* (1776), likewise, was one of the pioneering texts in the establishment of economics as a key social science.

Both Montesquieu and Ferguson emphasized that humans were social beings and recognized that sociality was an essential aspect of what they described as "human nature". As Ferguson wrote, human beings have "always wandered and settled, agreed or quarrelled in groups" (1995: 21). Humans, Montesquieu suggested, were flexible beings, able to adapt to different social circumstances, and the "desire to live in society" he considered a "natural law". Montesquieu's classic work was largely focussed on the nature and forms of government, but like Ferguson, he stressed not only the uniformity of human nature, but also the diversity of human cultures, whether in the form of tribes, nations, or empires. It was precisely the propensities of human nature that led, paradoxically, to "the diversity of human experience" (Montesquieu 1989: 3-7; Gay 1969: 339).

Universalism

The philosophes acknowledged both the uniformity of human nature and the fact that all humans were "by nature" a member of a community (Ferguson 1995: 59). They were therefore led to embrace the doctrine of *universalism* (or humanism). This doctrine implied a fundamental emphasis on the fact that all human beings belong to the same species, and therefore have a right, as humans, to autonomy, equality, and dignity independent of the race, gender, ethnic culture, religious tradition, or political nation to which they belonged. It put a stress on people's shared humanity. This is well attested by the fact that both Denis Diderot and Thomas Paine described themselves as "citizens of the world" (Gay 1966: 13). A distinction, therefore, has to be made between universal human rights and the political rights one holds as a citizen, although some scholars contend that all rights are of a political nature deriving from some state, and that the concepts of "liberty" and "equality" are simply ideological "fictions" emanating from the *State* and a cover for social inequality and political oppression (Douzinas 2010: 83).

But the Enlightenment thinkers clearly saw themselves as the "party of humanity", even though they were European men of property and their writings have a decidedly Eurocentric bias. Nevertheless, the philosophes, like the stoics, expressed a cosmopolitan sensibility, and Kant's famous categorical imperative implied that moral precepts were only valid if they could be "universalized"—that is, that they were applicable to all humans (Todorov 2009: 115).

Of course, it has to be recognized that during the eighteenth century a wealth of anthropological data was being gathered. This data was not only about Chinese, Indian, and Islamic civilizations, but also about tribal peoples, specifically, Native Americans and the people of the Pacific Ocean. So the philosophes were fully aware of the diversity of human cultures. This is evident from the writing of Montesquieu and Ferguson, as well as that of Rousseau, whose *Discourse on the Origin of Inequality* (1755) laid the foundations not only of romantic primitivism—well expressed today by John Zerzan (2008)—but also, according to Levi-Strauss (1976: 51), of anthropology. The philosophes often sympathetically wrote about other cultures—Diderot on the people of Tahiti, Rousseau on Native Americans—in order to advance a cultural and political critique of their own culture, eighteenth-century France. It was an early example of anthropology as cultural critique (Marcus and Fischer 1986).

But the emphasis on the diversity of human cultures—Ferguson described the multiplicity of social forms as "almost infinite" (1995: 62)—did not entail the espousal of cultural relativism, either in terms of moral or political values, or with regard to human knowledge.

The affirmation of humanism (universalism) was not viewed by the philosophes as antithetical to the recognition of the plurality cultures and identities—as it seems to be for many postmodernist scholars! In fact, as both Gay and Todorov stressed, for the Enlightenment thinkers these were closely interrelated (Gay 1969: 339; Todorov 2009: 122). Enlightenment thought expressed, in dialectical fashion, an advocacy of both unity and diversity, universalism and pluralism. Such thought, Todorov argued, cannot be conflated with either the dogmatic assertion of the supremacy of one's own culture, or with the nihilistic embrace of cultural relativism (2009: 123). Thus the Enlightenment placed a fundamental emphasis on humanity as a mode of subjectivity and as a source of moral and political values; it stressed humanness as a significant primary identity. For humanity as a universal concept is, along with personal identity (selfhood) and gender, one of the primary identities of all humans (Jenkins 2008: 74-84).

To suggest, as many postmodernist scholars and cultural pragmatics do (Baert 2005: 35), that universalism is a detached "view from nowhere", whether with regard to human knowledge (truth) or to morality, is quite misleading. For universalism implies a *human* earthly conception—not that of some deity!—which is quite distinct from both idealist metaphysics (such as that of Plato and Hegel) and from cultural relativism. Indeed, with respect to ethical theory, contemporary scholars have reaffirmed the universalism of the Enlightenment, and stressed that the recognition of cultural diversity (pluralism) is perfectly compatible with the acknowledgement of universally valid principles (Audi 2007; Hasan 2010).

Historical Naturalism

After outlining above the four key principles or concepts of the Enlightenment—the concept of reason, the politics of liberty, individualism, and universalism (humanism)—I want to turn now to the kind of metaphysics that emerged during the eighteenth century. This may be described as a form of historical, or evolutionary, materialism (Naturalism).

Although perhaps the majority of the Enlightenment philosophers were religious thinkers—Kant especially—the "spirit" of the Enlightenment expressed a very different metaphysic to that of religion (fideism), namely, an embryonic form of (historical) naturalism. Such naturalism implied both the advocacy of the scientific method and an embrace of ethical humanism—viewing morality as stemming not from some divine source, but from our existence as social animals with reason and language (as Aristotle had implied).

However, the philosophes had inherited from earlier classical science a rather mechanistic conception of nature, as expressed by both Cartesian philosophy and Newtonian physics. Mathematical physics was the key science. It expressed an emphasis on matter in motion, an all-embracing determinism, a subject-object dualism, and a stress on the experimental method. Both La Mettrie and D'Holbach embraced this mechanistic materialism. It entailed, however, not only a complete rejection of all forms of spiritualism, but also the downplaying of the living and of the social aspects of human existence.

But what is of crucial importance is that during the eighteenth century, with the emergence of such sciences as physiology, biology, palaeontology, geology, and anthropology (in the broadest sense) this conception of science (and of nature) began to be seriously undermined. That the world was essentially static and mechanical began to be questioned and the dualism between history (the human world) and nature became completely undermined. Science was no longer dominated by mechanical physics but began to incorporate a historical understanding, not only of human history (social life), but also of nature. Buffon's vast compilation of biological (and geological) knowledge was significantly entitled *Histoire naturelle* (*Natural History*) (1749-1766), and implied that nature had a history, even though Buffon himself, like Aristotle, and Carl Linnaeus, still believed that biological species were unchanging entities.

The recognition of the great antiquity of the earth and of the evolution of organic life—particularly emphasized by Diderot, Erasmus Darwin, and later by Jean-Baptist de Lamarck in his *Philosphie zoologique* (1809)—inaugurated a very different conception of nature and the emergence of the historical sciences. The Enlightenment thus initiated a form of historical naturalism, and eighteenth-century biologists such as Buffon have been described as "forerunners of Darwin" (Cassirer 1951: 77-79; Hampson 1968: 218-219; Outram 1995: 58-59).

The notion that the Enlightenment revolution wished to "abolish" or "eliminate" the past (as implied by Bruno Latour 1993: 69) is completely biased and misconceived. Although "historicism" may be described by a reactionary aristocrat like Nietzsche as a "disease", what is significant about the philosophes is that they explored, and introduced through historical understanding, both "deep time" and an evolutionary conception of the earth and organic life (but c.f. Gee 2000: 8 on whether palaeontology is, in fact, a historical science, given Gee's assumption that history and science are antithetical concepts).

In fact, time as history, the irreversible "arrow of time", has been viewed by many scholars as a key characteristic of the so-called "modern" episteme, especially as reflected in the studies of thermodynamics and of evolutionary theory (Foucault 1970: 219-220; Prigogine and Stengers 1984; Gould 1987; Latour 1993: 68-69).

With regard to metaphysics there is therefore an essential tension within the Enlightenment between mechanistic materialism and synchronic science (as well as deism), and historical naturalism.

Within the emerging social sciences, human social life and cultures were equally seen as historical phenomena, given that social forms changed through historical time. In his classic study, Adam Ferguson (1767) made one of the earliest attempts to outline a social evolutionary scheme, and clearly believed that the human species, like the human individual, was characterized by a progressive development. He noted the similarities between the social life of early Europeans and that of Native Americans, who he termed "savages" as they both lacked agriculture and the concept of property. But whether human life was progressing, or what "progress" actually entailed, was an issue around which the philosophes were deeply divided. Nevertheless, there was a belief among the philosophes that the development of secular knowledge, particularly the application of science, would lead to both material and moral improvement in the human condition. Sadly, given the history of the last two centuries—with the rise of the modern nation-state and the development of industrial capitalism, specifically machine technology—the hopes and ambitions of the philosophies never materialized, and the Enlightenment "project", as the postmodernists described it, has generally been depicted as a "failure". Nevertheless, many important scholars continue to affirm and advocate the "spirit" of the Enlightenment and the need to uphold its basic values, namely: a trust in human reason; a passion for free enquiry; the affirmation of equality and liberty as universal principles; the advocacy for historical naturalism as an ontology; and secular humanism as an ethic (Bookchin 1995B; Wilson 1998: 13-47; Bunge 1999: 129-143; Bronner 2004; Todorov 2009; Pagden 2013).

Needless to say, a number of intellectual disciplines and traditions all have their roots in the Enlightenment: not only anthropology (and the social sciences more generally) but also the radical political traditions of liberalism, feminism, socialism, and anarchism.

But three final reflections may be made on the Enlightenment as an intellectual tradition.

Firstly, many of the basic tenets of Enlightenment thought—the emphasis on reason; the positive evaluation of human praxis and empirical knowledge; and the enduring values of liberty, equality, and fraternity (social solidarity)—are by no means unique to eighteenth-century Europe. These tenets have been evident throughout human history, in many social contexts (Bronner 2004: 31; Todorov 2009: 130).

Indeed, the rallying cry of the French Revolution, with its emphasis on individual autonomy, egalitarian relations, mutual aid, and social solidarity, finds a clear echo in the social life and "anarchic solidarity" of many hunter-gatherers and swidden cultivators in many parts of the world (Gibson and Sillander 2011; Morris 2014B: 217-237, 2015).

Secondly, it is quite unhelpful and misleading to *equate* the Enlightenment with so-called "modernity" (still less with Western civilization) which seems to be a common tendency among postmodernist scholars. Modernity is usually defined in terms of an era, and with the social and political transformations of the nineteenth century, or even earlier. These developments consist of the following: the emergence of industrial capitalism, with the development of the productive forces in the form of an ever-expanding economy; the rise of the modern nation-state, with its centralization of political power, the advent of bureaucratic administration, and the promotion of state schooling (and thus the spread of literacy); the accompanying advocacy of the ideology of Nationalism and of a national identity; an increase in urbanization with the growth of industrial cities and the separation of various societal functions; and finally, the impact of the scientific revolution and the increasing secularization of philosophy, if not Western culture (Habermas 1987: 2-20; Patterson 1997; Bunge 1998: 271).

Depicted as an era, from at least the seventeenth century, "modernity" has thus been viewed in institutional terms and as comprised of various intertwined strands. To describe "modernity", as defined above, as having never happened (Latour 1993: 47) is to verge on obscurantism. It is typical of the anti-realist metaphysic that pervades Latour's entire work, in which he also informs us that neither human cultures nor nature exist (1993: 104).

Modernity, however, has been described not only as a historical era in terms of radical social transformations, but also as a "mode of thought". Almost every social theorist seems to have his or her own conception of what constitutes the culture of modernity. I have no wish to become entangled in the complex debates surround the nature of "modern ideology"; it will suffice to briefly mention the work of two influential scholars: Jean-Francois Lyotard and Louis Dumont.

Lyotard, for example, famously defined modernity in terms of the articulation of "meta-narratives". He had in mind, in particular, Hegel's idealist philosophy, which purported to outline the universal history of the world spirit (*geist*) as expressed through the human mind (*culture*); Marx's social theory, Lyotard having been in his early years a committed Marxist; and, finally—and especially—the radical Enlightenment. All three were viewed by Lyotard as "grand narratives" of human emancipation. The Enlightenment was thus misleadingly equated with modernity. Lyotard's book *The Postmodern Condition* (1984), was, in fact, virtually the founding text of postmodernist theory (Malpas 2003: 25-26).

Louis Dumont, in contrast, defined "modern ideology" in terms of three basic criteria, namely that human relationships with the natural world (science) were given primacy over social relationships (humanities); that an atomistic epistemology was favoured, as expressed in the ideology of individualism; and, finally, that a radical dualism was advocated between human life (culture) and nature. Modern culture was thus virtually equated with the Cartesian worldview (and positivism) (Dumont 1986: 36).

Other scholars, however, have defined modern culture as the exact *antithesis*of the Enlightenment (and Aristotle), equating it with Neo-Kantian philosophy and the downplaying of the importance of human reason (Hollinger 1986: 38-59).

Given its broad coverage and the rather abstruse ways in which it has been employed in recent scholarship, I personally, in my various writings, have never utilized the concept of "modernity". But it is, I think, nevertheless important to retain a clear distinction between the Enlightenment and modernity—the institutions and culture of the modern world, specifically industrial capitalism and the modern nation-state. In fact, the radical values expressed by the Enlightenment were usurped, realigned, and redefined by the various forces of "modernity".

Finally, the ideology of "progress" was certainly an important aspect of the Enlightenment tradition. The belief that the course of human history was one of continuous progress, and even entailed the "perfection of the human species"—as the Marquis de Condorcet enthusiastically put it (Kramnick 1995: 37)—was a crucial dimension of the Enlightenment. It is then hardly surprising that the Enlightenment, like Marxism, has been viewed as a secular version of the Christian ethic of salvation and envisaged the culmination of human history in the establishment of the kingdom of God on Earth.

But many scholars have long recognized that the Enlightenment was an ambiguous legacy, and that modernity was also a "mixed blessing". Following Peter Kropotkin, the social ecologist Murray Bookchin, has for example, emphasized that human "civilization", since its inception, has always consisted of two legacies or strands.

On the one hand, there is what Bookchin described as the *legacy of domination*, reflected specifically with the rise of industrial capitalism and the modern nation-state. This legacy entailed a growing concentration of economic power, and thus widespread social inequalities; a "dialectic of violence", reflected in the disintegration of local communities; the denial of human rights, and other forms of political oppression by tyrannical governments; the widespread existence of weapons of mass destruction and the very high levels of military expenditure throughout the world; and, finally, an "ecological crisis" reflected in the widespread pollution of the oceans and atmosphere, the negative impact of industrial farming, deforestation, the creation of toxic wastelands, and a serious loss of biodiversity (Morris 2004B: 15-17). Bookchin concluded that "The failures of civilization have been enormous and have claimed a ghastly toll of life." (1999: 130).

On the other hand, there is within human civilization a *legacy of freedom*, for unlike nihilistic postmodernists and anarcho-primitivists, Bookchin was adamant that civilization has its positive aspects. Significantly, Bookchin identified the legacy of freedom with the Enlightenment, which he described as a "glorious project". It implied, he argued, through the employment of reason and empirical science, a much deeper understanding of the natural world in which we find ourselves, and therefore an overall improvement, however ambiguous, of humanity's material conditions of life. It allowed a more democratic form of politics and the emergence of a rational form of ethics—ethical naturalism—that was more progressive than unreflective custom and the dogmatic ethics of most religious traditions. it also entailed the widespread commitment to such universal values as individual autonomy, reason, religious tolerance, the promotion and protection of human rights and civil liberties, and the value of education and cultural diversity (Bookchin 1982, 1995B; Grayling 2009: 111-116; Morris 2012: 252-256).

As Jonathan Israel concluded, the Enlightenment "has been, and remains by far the most positive factor in shaping contemporary reality and those strands of "modernity" that anyone wishing to live in accord with reason would want to support and contribute to" (2006: preface).

Counter-Enlightenment

As noted earlier, postmodernist scholars, or cultural theorists, continue to express their disdain and to denigrate the Enlightenment by suggesting that its humanistic ideals (universal values) simply provided the moral basis for colonial domination and totalitarian politics, or even worse, the Holocaust. This portrayal of the Enlightenment is misleading and biased—to say the least.

Of course, it has long been recognized that theories of sociocultural evolution, and Darwinian theory more generally, have been employed in Eurocentric fashion to denigrate non-Western people, especially tribal peoples, or to bolster imperial rule. Kropotkin long ago emphasized how the Darwinian conception of natural selection and the idea of "survival of the fittest" had been used to justify laissez-faire capitalism and colonial expansion (Morris 2004B: 134). But, as Bruce Trigger has discussed, there is no intrinsic connection that links Darwin's evolutionary biology and other specific theories of sociocultural evolution (as initiated by Enlightenment thinkers like Adam Ferguson) with either free-market capitalism or colonial rule (1998: 3-8). Nor is there any intrinsic connection between universalism—the expression of human values and objective truths—and colonialism, doctrines of white supremacy, or totalitarian politics.

It also has to be recognized that there is no aspect of human life—whether universal values such as freedom, equality, the arts, sports, festivals, ideas of democracy, or various aspects of science—that has not been appropriated by clerics, industrial capitalists, and political tyrants—that "dark trinity" as Ricardo Flores Magon described them—to affirm ecclesiastical authority, to further economic exploitation, or to bolster political oppression. It is an interesting question why the Enlightenment is singled out for contempt and denigration by culture theorists and the postmodernists in their various guises (post-colonial, post-anarchist, post-human) rather than the arts and the political ideologies of the twentieth century. It is significant that the Mexican anarchist Flores Magon, in contrast, fervently supported the Enlightenment (Bufe and Verter 2005).

It is therefore rather biased, and totally misleading, to interpret the Enlightenment and the concept of reason as implying a form of totalitarian thought. This negative view was well expressed by the critical theorists (Marxists) Theodor Adorno, and Max Horkheimer in their renowned, and highly influential study *Dialectic of Enlightenment* (1973). It was written during a period of extreme pessimism among many scholars, in the aftermath of the horrors of the Second World War. These scholars implied that the "concept" of the Enlightenment involved the following: an instrumental form of rationality that entailed the technological mastery of nature and the "dictatorship"—the total administration of humans; the "disenchantment of the world" with the rejection of animistic and religious beliefs, and thus the loss of any "claim to meaning"; a positivist vision that radically separated knowledge from human values; and, finally, the equation of the Enlightenment with the "culture industry", interpreting it as "mass deception".

It is, however, important to note that they intended their study as preparing the way "for a more positive notion of Enlightenment which will release it from the entanglement with blind domination" (1973: xvi). In fact, what their book describes is not the Enlightenment (as normally understood) but the ideology of instrumental reason, and, as they put it, the "conversion of Enlightenment into positivism" (x). Thus what Adorno and Horkheimer eventually portray is the "betrayal" of the ideals and essential tenets of the Enlightenment under the ethos of industrial capitalism and the growing powers of coercive nation-states—modernity.

Through the "programme" of the Enlightenment, with its implicit advocacy of evolutionary naturalism, may have implied or enhanced the "disenchantment" of natural phenomena, in fact Max Weber's famous thesis on the "disenchantment of the world"—derived from the German poet Johann Schiller—had little to do with the Enlightenment per se. For this disenchantment was specifically related to the monotheistic religions Christianity, and Islam—the two religions of the book which were primarily responsible for the elimination of magic and the decline of tribal animism (paganism) within Western culture (Geryll and Mills 1948: 139; Robbins 2011).

But, of course, the "disenchantment" of the world—the fact that the earth no longer has a religious significance—did not entail that the natural world ceased to have meaning and significance for humans; historical naturalism and evolutionary theory affirmed the contrary. Meaning in life stems from people's interactions with the natural world and is largely independent of both language and religious ideology (Hoffmeyer 1996).

The notion that the Enlightenment was responsible for the atrocities of the twentieth century—two world wars, the Bolshevik tyranny, the rise of fascism, and even the Holocaust—is philosophically untenable, and historically rather bizarre. Such a theory, trumpeted in oracular fashion by the postmodern acolytes of Adorno and Horkheimer, completely neglects to engage with many crucial historical factors: inter-state conflicts, fascist ideologies that completely denounced Enlightenment universalism and all it stood for, and the realities of a rapacious, ever-expanding global capitalism. It is rather like blaming Jesus of Nazareth for the Catholic Inquisition and the witch "craze" of the seventeenth century.

It is equally misleading to equate science and reason with the domination of humans and nature. Significantly, Bronner affirms that neither Adorno nor Horkheimer had any real appreciation of the democratic inheritance and emancipatory potential of the Enlightenment (Bronner 2004: 1-16).

Needless to say, there was no real connection between colonial states and the Enlightenment—whether in terms of its evolutionary naturalism (atheism), its secular humanism (ethical naturalism), or its democratic politics. Contrary to post-colonial critiques of the Enlightenment, colonial states abetted and supported the world religions, whether Islam or Hinduism, and actively promoted Christianity, particularly in sub-Saharan Africa. Colonial states were fervent in promoting the *counter*-Enlightenment, namely religious faith, social hierarchy, and political authority. Colonial states, like many contemporary states, promoted and combined the three constituent elements of Flores Magon's "dark trinity": a religious metaphysic (spiritualism), free-market capitalism (commerce), and authoritarian politics.

Two points are worth making with respect to contemporary politics. Firstly, as Richard Dawkins noted, given the rise of evangelical Christianity and "creationist" theology in the United States, it is now "virtually impossible for an honest atheist to win public election in America" (2006: 45)—that is, anyone who affirms the evolutionary naturalism of the Enlightenment. Secondly, the writings of Meera Nanda, in defence of the Enlightenment as a cosmopolitan tradition, have indicated that there is a close—almost unholy!—alliance between the anti-Enlightenment and the anti-science critiques of the Indian post-colonial theorists (such as Ashis Nandy and Claude Alvares), namely in the promotion of "Vedic science" and extreme right-wing Hindu nationalism (Nanda 2003).

Critiques of the Enlightenment are, of course, nothing new. Reactions to its materialist ontology, social ethics, and democratic politics go back to the end of the eighteenth century and the French Revolution. The reaction is generally known as the counter-Enlightenment. It thus has to be recognized that the "making of totalitarian thought", as Josef Llobera (2003) cogently argued, has less to do with the "spirit" of the Enlightenment than with its exact *antithesis*—the counter-Enlightenment. For in the aftermath of the French Revolution—its "failure" is often viewed as marking the end of the Enlightenment (Berlin 1979: 24; Morris 1996: 93-107)—there was an acute social and political crisis throughout Europe. The outcome of this crisis was a conservative reaction—the counter-Enlightenment—in the early decades of the nineteenth century. Much has been written on this movement, which was particularly associated with such scholars as Edmund Burke, Joseph de Maistre, Louis Bonald, Joseph Gobineau, and Johann Hamann. What they represented was an extreme conservative reaction to the Enlightenment. Whereas the Enlightenment valued the critical exercise of reason and put an emphasis on liberty and individual rights, the counter-Enlightenment in contrast stood for coercive authority, obedience, and tradition (Bronner 2004: 67). Whereas the Enlightenment placed an emphasis on evolutionary naturalism, scientific reason and a cosmopolitan sensibility—a common humanity—and individual autonomy, the counter-Enlightenment stressed religious faith, mystical intuition, ethnic nationalism and racism, and the absolute power of the social—whether the church or state—over the individual.

The notion that cosmopolitanism implies "the shredding of all that makes one human", as Isaiah Berlin interprets it (1979: 12), thus denying the social aspects of human life, is quite misleading. There is no antithesis, as Enlightenment thinkers realized (but postmodernists do not!) between our shared identity as humans—our humanity—and the fact that all humans, in all societies, have multiple or plural identities. These identities relate, for example, to gender, kinship, ethnicity, class, occupation, and to political and religious affiliations, as well, of course, as to the fact that people have a unique or personal identity (selfhood) (Morris 1994: 1-22; Sen 2006; Jenkins 2008; on the counter-Enlightenment see especially Nisbet 1986; Berlin 1990).

But critiques of the Enlightenment continue to be recycled. Two examples will suffice: one from a post-anarchist, the other from a post-human scholar.

In a well-known essay on the "epistemological basis" of anarchism, the self-styled post-anarchist Andrew Koch, inspired by Nietzsche, offers as a representation—while rejecting the very idea of "representation"!—of William Godwin's, Peter Kropotkin's, and Pierre-Joseph Proudhon's conceptions of the human subject. Godwin, of course, was a key figure in the British Enlightenment, a passionate rationalist and individualist. Koch comes to the rather obvious conclusion that all these three anarchists posited a "universal human identity". Who doesn't? Apart from racists and extreme egoists? Koch tells us that the human subject is a blank slate, an organism but devoid of any biological, psychological, or social attributes. He then recites the familiar postmodern mantra that "universalism" entails "totalitarian politics" (2011: 36). Koch has things completely back-to-front. Radical anti-humanists and anti-universalists like, for example, de Maistre and Adolf Hitler, were racists and advocates of totalitarian politics. In contrast, an Enlightenment universalist like Godwin—and contemporary radical scholars like Noam Chomsky and Murray Bookchin—were anarchists and advocates of anti-state libertarian politics.

Enlightenment humanism, according to one self-styled "post human" [sic] scholar, was not only androcentric, but entailed the following four tenets: it implied a Cartesian unitary subject that failed to acknowledge that humans were social beings with multiple identities; it viewed humans as transcendental rational egos, ignored the fact that humans were a part of nature, and thus neglected the importance of empirical knowledge; it advocated a racist doctrine of white supremacy that masqueraded under an ethic of universalism and equality; and, finally, it postulated a radical dichotomy between human social life (culture) and the natural world (nature), thus failing to recognize the essential continuity between nature and culture (Braidotti 2013).

This "representation" [sic] is a travesty of Enlightenment humanism and can only be described as sterile, biased, and completely lacking any historical sensibility. It hardly does justice to the complexity and innovative nature of the "spirit" of the Enlightenment (discussed above), which is well portrayed by such historians as Jonathan Israel (2001) and Anthony Pagden (2013).

It was, of course, during the Enlightenment that the philosophes, especially D'Holbach and Diderot—drawing on Spinoza—reaffirmed and established a materialist ontology, a metaphysic now embraced by Braidotti! The social nature of the human subject was also highlighted by the philosophes (Montesquieu, Ferguson, Herder, Adam Smith), as we discussed above. This complemented the stress on individual autonomy—also expressed by Enlightenment thinkers—and thus not only undermined the Cartesian unitary subject but led to the establishment of anthropology, history, and sociology as key disciplines in understanding human subjectivity. Equally important, as I also emphasized above, the Enlightenment philosophes (such as Baffon, Diderot, and Lamarck) completely undermined the Cartesian dualistic metaphysic—the radical opposition between humans and nature—in developing palaeontology and evolutionary biology. The Enlightenment humanists therefore recognized, long before Braidotti (2013: 82) that human subjectivity was "embodied" (see Porter 2003) and "embedded" in both the natural world and within a socio-historical context—that humans were an intrinsic part of nature and that there was therefore an essential "nature-culture continuum".

Braidotti writes that humanistic universalism is distinctive of Western culture, its own specific form of "particularism" (2013: 20).

This is a highly Eurocentric viewpoint, not only failing to recognize the cosmopolitan and global impact of the Enlightenment, but also the fact that humanism (universalism) is a philosophy of life that is found across the world and throughout human history. All humans recognize and respect such basic universal values as life, health, friendship, loyalty, reciprocity, fairness, knowledge, peace, and human liberty (Bunge 2001: 11)—like Buddha, Confucius, and Spinoza, as well as the Enlightenment philosophes, were all humanistic scholars. It is ironic that Braidotti, following her mentor Gilles Deleuze, continually extols Spinoza's materialist worldview (2013: 56), yet is completely unaware that Spinoza was a key figure in the development of the radical Enlightenment (Israel 2001), which Braidotti consistently denigrates through misrepresentation and caricature.

The postmodernists, following either Richard Rorty or Gilles Deleuze, reject the very idea of "representation"; they offer instead, like Braidotti, a complete *misrepresentation* of the Enlightenment.

Needless to say, contra Braidotti, none of the well-known humanist scholars of the twentieth century—for example, Lev Vygotsky, Ernst Cassirer, Erich Fromm, Lewis Mumford, Rene Dubos, Maurice Merleau-Ponty, Murray Bookchin, and Paul Kurtz—were "humanists" according to Braidotti's own bizarre definition. Yet all, in various ways, initiated a form of enlightened (or ecological) humanism that long ago anticipated Braidotti's own concept of "nomadic" subjectivity (Morris 2012, 2014A). Is this "nomadic subject" a *universal* category or merely a description of a globe-trotting academic?

The liberal scholar Isaiah Berlin compellingly summed up the virtues of the Enlightenment: "The intellectual power, honesty, lucidity, courage, and disinterested love of truth of the most gifted thinkers of the eighteenth century remain to this day without parallel. Their age is one of the best and the most hopeful episodes in the life of mankind." (1956: 29).

For the Enlightenment advocated a materialist ontology and an embryonic form of evolutionary naturalism that completely undermined Cartesian metaphysics: its anthropocentrism, its mechanistic conception of nature, and its dualistic worldview. Matter was thus viewed as sensitive and self-moving, and the Enlightenment thinkers, long before Braidotti (!) emphasized the essential continuity between nature, human social life, and culture.

This gave rise to an ethical naturalism that completely undermined the entire structure of a divinely ordained morality. The political philosophy of the Enlightenment emphasized both the autonomy of the individual and equality; the universality of such human values as religious tolerance, fairness, free enquiry, and democracy, thus effectively undermining all *legitimation* for monarchy and the divine rights of kings, gender inequalities, aristocratic rule, and slavery. We must therefore hold fast to the ideals and the "spirit" of the radical Enlightenment (Israel 2000; Bronner 2004; Todorov 2009; Pagden 2013).

I shall turn now to the two key tenets of evolutionary (dialectical) naturalism: a realist ontology, and a materialist metaphysic (or worldview).

Chapter Four

Ontological Realism

Apart from religious mystics, idealist philosophers, and postmodernist anthropologists, everyone in the world recognizes that there is a real, material world that exists independently of us humans. For example, when tribal foragers in South India are confronted with severe inclement weather, a lack of food (especially honey), or chronic ailments, they appeal to the ancestral spirits or the mountain deities (mala devi) for support and for the relief of their suffering (Morris 2014B: 226).

But during the latter part of the last century, it became quite fashionable among many philosophers and social scientists to theorize that the natural world, along with its biodiversity—or even "reality" itself—was a *social* construct! In this chapter I shall offer some critical reflections on this anti-realism, as well as outlining and advocating ontological realism as an essential premise for any understanding of the natural world and human social life. A critique of anti-realism is, in fact, a recurrent theme in this manifesto and will be discussed again in chapter eight.

Reflections on Anti-Realism

It has, of course, long been known—long before postmodernism came upon the anthropological scene—that we do not perceive or experience the world in pristine fashion. For our engagement with the world is always mediated by our personal interests, by our state of mind, by language and cultural conceptions, and, above all, by social practices. As an early and important historian of science, William Whewell, put it in 1838: "There is a mask of theory over the whole face of nature." (Megill 1994: 66). Scholars as different as Lewis Mumford and Ruth Benedict long ago emphasized that sociocultural factors influence, modify, or "edit" how humans perceive the world—even that they do not wholly determine such relations (Benedict 1934: 2; Mumford 1951: 23).

But as with Wilhelm Dilthey, Benedict's important mentor, this insight or affirmation did not in the least imply a denial of the reality of the material world. This important insight, which has been part of the common currency of the social sciences ever since Marx, was taken up with some fervour by many philosophers, literary theorists, and postmodernist scholars towards the end of the last century. But they seem to have taken this important insight to extremes, and in a "veritable epidemic" of "social constructionism" and "world-making" propounded a latter-day version of Kantian transcendental idealism. They went even further than Kant, denying the reality of the material world—the things-in-themselves—as well as ignoring the importance of practical (empirical) knowledge. Kant, as a physical geographer, of course never doubted the reality of the material world. He simply held that its real nature was unknowable, given that the objective world was a construct from subjective experience, as Whitehead put it (1929: 181).

Cultural idealism, in its various guises, thus became all the rage in the halls of academia, and was embraced by a wide range of scholars (e.g., Goodman 1978; Rorty 1980, 1989). Such constructionism combines two basic Kantian ideas: that the world as we know it is constituted by our concepts, and that an independent world is forever beyond our ken (Devitt 1984: ix). But, as already mentioned, many scholars went even further in an anti-realist direction, denying the independent existence of a world beyond our cognition—a world of things that have causal powers and efficacy. With the free use of the term "worlds" they invariably conflate the cognitive reality which is culture—"discourses" is now the more popular term—and the material world that exists independent of humans. Thus anthropologists now tell us that there is "no nature, no culture", or that nature (including bacteria and tigers), sex, emotions, and the body are purely social constructs, or even that they do not "exist" outside of Western discourses or "theory".

In a rather confusing analysis that presupposes a distinction between the natural world and human culture—specifically "spiritual worldviews"—one anthropologist informs us that with any perception of the natural world, nature "becomes" culture. The anthology on indigenous (tribal) knowledge, specifically relating to trees and forests, is thus appropriately entitled "Nature *is* Culture" (Seeland 1997: 103).

Even Marxists seem to have embraced this Neo-Kantian cultural idealism, for two well-known structural Marxists write that "Objects of discourse do not exist. The entities the discourse refers to are constituted in it and by it." (Hindness and Hirst 1977: 20). A classic and typical example will suffice to illustrate the widespread "anti-realism" that developed within anthropology and the social sciences at the end of the last century.

In a well-known and pioneering ethnographic study on *"Laboratory Life"* (1979), significantly subtitled "the construction of scientific facts", Bruno Latour and Steve Woolgar declare that the "out-there-ness" (i.e., the external world!) is "the *consequence* of the scientific work rather that its *cause*", and go on to suggest that "the thing and the statement correspond for the simple reason that they come from the same source. Their separation is only the final stage in the process of their *construction*." (1979: 182-183).

Although they did not wish to suggest that facts did not exist, or that there is no reality, Latour and Woolgar nevertheless write: "Despite the fact that our scientists held the belief that the inscriptions could be representations or indicators of some entity with an independent existence "out there" we have argued that such entities were constituted solely through the use of these inscriptions." (1979: 128).

Thus the external world and scientific facts are not independent of the inquiring subject: they are the product of theoretical "construction". Nothing could be more anthropocentric! The limitations of their approach are self-evident if we apply this same "constructionist" approach to their own social science, as expressed in their ethnography. Is the Salk Institute, as a scientific fact, merely a "fiction" or "artefact" of their own "literary inscriptions"? Woolgar was later to affirm that the mode of analysis that he and Latour shared was "consistent with the position of the idealist wing of ethnomethodology, that there is no reality independent of the words (texts, signs, documents, and so on) used to apprehend it. In other words, reality is constituted in and through discourse" (1986: 312). This form of textualism (or cultural idealism) which reduces the material world and social life to texts (discourses) has been the subject of extensive critique (Gross and Levitt 1998: 42-70; Bunge 1999: 174-179; Bourdieu 2004: 26-30; Feist 2006; Searle 2009).

An iconoclastic thinker, Bruno Latour continually expressed anti-realist views, claiming for instance that human beings not only "construct" human collectivities but also the "non-humans" (mosquitoes, elephants?) that surround them (1993: 106). He also argues, like Rorty, that scientific disputes are settled by consensus (or power struggles!) and that the truth of scientific theories has no essential relationship to the material world—nature—that exists independently of humans. He even ridicules the "appeal" to nature as a way of resolving scientific controversies (Latour 1987: 94-100; Sokal and Bricmont 1999: 85-90). Likewise, as an avid social constructionist, Latour (1988) suggests that bacteria did not exist until discovered by Koch and Pasteur in the nineteenth century! This is a debatable, if not completely vacuous thesis (Bunge 2006: 66).

It is hardly surprising therefore that Latour has been interpreted by many critics (e.g., Sokal and Bricmont 1999) as an early advocate of postmodernism and epistemic relativism. But even in his early writings, describing himself as "non-modern", Latour, in self-contradiction, emphasised that there was "indeed a nature that we have not made" and that there are "indeed indisputable scientific facts" (1993: 140). He also in the same text made some telling critiques of postmodernism—identifying this intellectual movement with Baudrillard and Lyotard—describing it in highly negative terms as "nonsense" (1993: 115). He was later to repudiate his early advocacy of textualism and postmodernism, or, as he put it, he came to disagree with "most" of what he had earlier written (Latour et al 2011: 42), while continuing to berate and ridicule postmodernist scholars like Lyotard for having lost touch with reality—the real world! (Latour 2004B).

Going, like many postmodern anthropologists, from one extreme to another, Latour now embraces a form of atomism or reductive "materialism"—the actor-network or "affect" theory—derived essentially from outdated positivist sociology (Latour 2005). For Latour, God, dragons, witches, and Buffy the Vampire Slayer are now all treated as "actants". But, of course, as cultural concepts, they have no energy, do not change, and have no agency. Latour seems to have forgotten what Lucretius taught us long ago, namely, that "nothing can act or be acted upon, unless it is corporeal" (440, 1969: 14). A witch or Mickey Mouse, as cultural concepts, not only have no agency but also have no meaning apart from real human beings and a specific social context. Latour's actor-network theory is a cosmic version of the methodological individualism associated with the positivists Harold Garfinkel (1967) and Gabriel Tarde (2000). The "mechamosphere" of Deleuze and Guattani (1977) seems to form the appropriate backcloth.

Thus anthropologists like Latour, once besotted with language and power, have now come to recognize that humans are living beings and that their material encounters with the world involve many "active" things that are "non-human". Such old-fashioned reductive materialism is now heralded as a new, innovative approach to our understanding of social life and what it means "to be human"!

To suggest that "nature" (excuse the inverted commas!) is a construct or a "literary inscription"—rather than as something real, with respect to which we are continually engaged—or that nature *is* culture, are each highly problematic notions. One possible meaning of "nature" refers to the existential world in which we find ourselves: the trees, the clouds, the mountains, the animals and plants, the rocks, and all those natural processes which are independent of human cognition and on which human life depends. It is highly anthropocentric, if not plain absurd, to suggest that this is a human invention, creation, or artefact, or that this does not exist independent of cultural cognition or language (c.f. Dwyer 1996).On the other hand, one can simply mean by "nature" the highly variable cultural conceptions of nature. To suggest that these are social constructs is rather banal, though the suggestion is dressed up as if it was some profound insight.

Nobody, of course, has ever doubted the truth of social constructionism in the obvious sense that all cultural ideas, cultural practices, and social institutions are social constructions. Thus, though they may involve material entities and relations, they cannot be understood in purely biological terms (Bunge 2006: 79). That we are born into a world where "conceptual frameworks" exist is also hardly news or a contentious issue. Sociocultural factors do influence or modify the way we understand nature (in all its diverse aspects), as well as influencing the body, the emotions, and conceptions of ourselves as human beings. Anthropologists and social scientists have long recognized this—long before postmodernists and literary theorists. But "discourses" or "culture" do not *produce* the *objects* of our knowledge—as aspects of the natural and social worlds in which we live—and the suggestion that "our ways of understanding the world do not come from objective reality", but only from other people (Burr 1995: 7), is completely one-sided and misleading. Our knowledge of the world comes from three essential sources: innate pre-dispositions, our practical engagements with a material world that exists independently of us, and human social and cultural practices (Bloch 2012: 14). Human beings collectively create the cultural "worldviews" or "conceptual frameworks" by means of which we interpret and relate to the material world, but they only influence our understandings. They do not create the world as such.

Equally important, such cultural (cosmological) frameworks are completely distinct from our basic empirical or practical knowledge of the world, which derives not simply from other humans via language but from our *interactions* with the material world, as well as, to some extent, from our innate pre-dispositions, basic needs, and desires (Bloch 2012: 60-64). As one scholar insists: "human practices are socially constructed; bananas are not—for bananas exist independently of our concepts and cultural beliefs about them!" (Manicas 2006: 31; for a useful discussion of social constructionism see Hacking 1999).

In reacting against naïve realism, the notion that there is an isomorphic—a reflective or mimetic relationship between consciousness (or language) and the material world, per the early Wittgenstein's (1922) picture theory of language—postmodernist scholars and many anthropologists seem to have embraced (as I have indicated) a form of cultural (or linguistic) idealism. They have thus tended either to completely repudiate the notion of "representation", denying that language (in its many forms) has any relationship with the material world and human life experiences, or to deny the very reality of the material world (or nature) along with the body, sex, and emotions. Even the human subject has been de-materialized, for some literary theorists purporting to describe Pierre Bourdieu's sociology have absurdly proclaimed that languages (sign systems) not only "produce"*people* but also their thoughts, desires, and activities (Webb et al 2002: 33).

At this juncture it is worth noting that our knowledge of the material world—whether expressed in everyday empirical descriptions, cultural worldviews, or scientific theories—is never mimetic, or as Rorty (1980) misleadingly described it, a "mirror of nature". To the contrary, human knowledge of the material world, both natural and social, is always focussed and *cartographic* (symbolic), relating only to specific and limited aspects of the complex world in which humans find themselves. Knowledge is also, as Dewey emphasized, something that is *produced* in our relationship with the world and with other people. I shall return to the issue of "representation" in chapter ten.

Realism, interpreted here as implying a materialist outlook, is, as many scholars have insisted, a metaphysical (or ontological) doctrine. It is about what exists in the world and how the world is constituted. Contrary to what Kitten Hast writes, it is not a theory of knowledge, nor of truth, and it does not aim at providing a "faithful reflection of the world" (1995: 60). Realism as a doctrine is thus separate from semantic issues relating to truth and reference, and from issues relating to our knowledge of the world (epistemology)—both human and natural. Roy Bhaskar has critiqued what he describes as the "epistemic fallacy", the idea that ontological issues can be reduced to, or analysed in terms of, statements about knowledge—epistemology (1975: 36).

Ontological realism at its most basic is the notion or presupposition that there are real, concrete entities in the world, and that this external world exists independently of the knowing subject—independently of our experiences, our thoughts, our language, and our culture (Searle 1999: 10; Bunge 2006: 29). As an ontology, realism has been succinctly described as "the theory that there is an objective material world that exists independently of consciousness, and which is knowable by consciousness" (Sayers 1985: 3).

Realism thus implies a distinction—not a dualism—between the human subject and objects in the world. This distinction is enshrined in naïve realism, the tacit epistemology of nearly everyone. Although some trendy postmodernist anthropologists nonchalantly dismiss the subject/object distinction, it is clearly basic not only to humans, but to all forms of animal life. As Bunge put it: "Just think of the chances a gazelle would stand if it did not instinctively acknowledge real existence of lions in its outer world." (2006: 34). The very idea of an organism and its environment pre-supposes the subject-object distinction, and no living being would survive without some sense of subjective agency.

Everyone, of course, is a "realist" or a "fundamentalist" in some sense, making ontological assumptions about what is real and what "exists", metaphysics is thus not something that one can dispense with, or put an "end" to—as both positivists and Heidegger and his acolytes suggest. Often, of course, the term "metaphysics" is used simply to describe the idealist philosophies of either Plato or Descartes. For Plato "ideas", the universals that exit in a spiritual realm, were "real"; for Descartes, the transcendental ego was "real"; for some eco-feminists the mother-goddess is "real", while for empiricists the "real" is sense-impressions. As used here, and defined above, realism entails the view that material things exist independent of human sense experience and cognition. The independence and reality of the material world is experienced whenever a volcano erupts, or we mistake a stick for a snake, or we get lost in the woods. Getting things wrong or having our expectations confounded justifies our belief that the world does indeed exist independently of our cognitions about it (Sayer 2000: 2).

It is, I think, somewhat ironic—indeed surprising—that two largely forgotten, but insightful, philosophers, Alfred North Whitehead and John MacMurray, though both fundamentally religious thinkers, should also strongly affirm a realist ontology.

Describing his own metaphysic as the "philosophy of organism", Whitehead was deeply opposed to both empiricism and to the mechanistic science of the sixteenth century. By "organism" Whitehead meant such enduring things as electrons, individual molecules, cells, and all living beings—such as plants and animals (including humans)—as well as the earth, the mountains, the Egyptian pyramids, and Edinburgh Castle. He defined these "organisms" as a "nexus" of transient occasions—events.

But the potentiality and limits of such events—which he termed, confusingly, as "actual entities"—were "governed", he wrote, by its "datum": the "stubborn facts" and the "givenness" of the actual world. And our basic and primary perception of this world, Whitehead suggests, was one of "causal efficacy", our visceral feelings of always being in contact with a material world (1929: 130-151). Significantly, for Whitehead there are within the "inescapable flux" of human experience always some things which abide—the life-histories of enduring material things—the organisms (1929: 206).

Like Whitehead, the existentialist philosopher John MacMurray was highly critical of mechanistic science, particularly the degree to which it oblated the human personality. But he never questioned the importance of human reason, and always acknowledged the objective existence of a material world. It is worth quoting an extract from MacMurray's writings to illustrate what he meant by objectively—in relation both to the material world and to other humans.

It is our nature to apprehend and enjoy a world that is outside ourselves, to live in communion with a world which is independent of us.... And when we are completely ourselves we live by that knowledge and appreciation of what is not ourselves, and so in communion with other beings. That is what I term our objectivity, and it is the essence of our "human nature". (MacMurray 1932: 178)

MacMurray thus affirmed the integrity of the human organism and the essential paradox of human experience: that the material world is completely independent of humans—our objectivity (realism)—and that we are always in "communion" with it—our subjective experience.

But, of course, experience is not a property of the human subject, as phenomenologists and existentialist anthropologists tend to think. It refers rather to our *interactions* with a material world. This is how John Dewey long ago defined experience. For Dewey experience did not consist simply of sense impressions nor of subjective states, but rather of everything that happens between a human organism, as a unique agent, and its environment, both natural and social (Dewey 1925: 87-89; Morris 2014A: 152).

Metaphysical Realism

Ontological realism affirms the independence and objectivity of the material world. This does not imply that humans are independent of this world, for humans are an intrinsic part of nature; humans and nature are therefore dialectically interrelated. Realism is not therefore to be confused nor conflated, as many hermeneutic scholars and postmodernist anthropologists invariably do, with Cartesian dualistic metaphysics. Realism is not only distinct from this metaphysic but also from its reductive offsprings via Kant—cultural idealism and positivism. It is equally facile to suggest that realism posits a static universe, a world of unchanging, fixed objects, bereft of relations. Realism posits a world not only independent of humans, but one that is complex, structured, relational, and continually changing. Only idealists like Bergson, an inverse Platonist, conceives the world that we actually experience as consisting of inert, unchanging objects.

A thoroughly dualistic thinker, Henri Bergson described the world as consisting of two separate realms: that of inert matter, identified with space and understood through the intellect (science and common sense i.e. practical knowledge); and that of life (*elan vital*), identified with time as an undifferentiated flux (duration) and understood through spiritual intuition (idealist philosophy). As Dewey remarked, the flux (flow) of life was thus spiritualized and virtually equated by Bergson with the deity (Bergson 1907; Dewey 1925: 71-72).

Realism, which is intrinsically linked with philosophical materialism—Mario Bunge (2006: 33) coined the term *hylorealism* to express their intimate connection—is thus to be clearly distinguished from both idealism and positivism.

Idealism takes many different forms; it is essentially a family of doctrines that assert the independent existence or the primacy of *ideas*. Objective idealists (Plato, Leibniz, Hegel, Whitehead) view the world as if it was a spiritual being. They claim that ideas (spirit) exist objectively, with ideas, according to Plato, existing autonomously in a spiritual realm. This implies that the material world has no reality independent of the spiritual realm, or that it is simply the emanation or creation of spirit (or some deity). Objective idealism is thus a philosophy of transcendence, and the laws of nature derive from, as Whitehead put it, "a transcendent imposing deity" (1933: 118).

Subjective idealism (Berkely, Kant), on the other hand, is akin to phenomenalism in denying that the objective material world exists independently of subjective experience. However, during the twentieth century many forms of subjective or cultural idealism have emerged, namely, phenomenology (Husserl), Neo-Kantian philosophy (Cassirer), hermeneutics (Heidegger), social constructionism (Goodman/Latour), as well as textualism and postmodernism (Rorty, Derrida, Tyler). According to Mario Bunge all these forms of idealism express an anti-realist metaphysic and a "retreat from reality" (2006: 56-87). I discuss such Neo-Kantian tendencies more fully in chapter eight below.

It is worth noting, in addition, that all forms of religion (spiritualism) essentially imply an idealist metaphysic—whether taking the form of animism, theism, or panentheism, and whether or not involving beliefs in magic, shamanism, deities, ancestral spirits, witches, or immortal souls. Idealism, in summary, entails the view that the material world either has no reality, or is the emanation of spirit (or some deity), or that external things do not exist apart from our knowledge or consciousness of them.

Positivism, in contrast, firmly equates the real world with phenomenon (what is experienced) and thus embraces what has been described as phenomenalism or empirical realism. David Hume, Ernst Mach, and Richard Rorty are well-known exemplars of this approach. Although usually acknowledging—like Kant—that the material world exists independently of the human subject, it denies any distinction between appearance and reality, i.e., between the world as it appears to us as humans and the world as it is in reality. Mario Bunge rather derisively suggests that a positivist or phenomenalist philosopher such as Rorty (1998: 73) is rather like a second-hand car dealer who assures us that what we get is what we see (2006: 85; Malik 2000: 343-345). Unlike realism, positivism denies therefore the reality of unobservables—whether entities (like bacteria), structures (relations), or historical processes—and thus rejects explanations in terms of underlying generative structures or causal mechanisms (Collier 1994: 26-27; Bunge 1999: 29).

In his discussion of the laws of nature, Whitehead stressed that positivism simply recorded correlations or successions of observed facts. As a form of explanation, it was therefore merely *descriptive* (1933: 199-121; Bunge 1999: 29).

Outside of some philosophy departments, or among religious mystics as well as among some cultural or postmodernist anthropologists, ontological realism is universally held by everyone. It forms the basis of common-sense understanding—practical knowledge, or "peoples' science" as Paul Richards describes it (1985: 155)—and the empirical sciences. Common sense, of course, *sensus communis*, can be interpreted in Aristotelian fashion as a kind of sixth sense that draws together the localized senses of sight, touch, smell, taste, and hearing. It is this sixth sense, Hannah Arendt writes, that gives us a sense of realness regarding the world (1978: 49). Like Arendt, Karl Popper critically affirmed the importance of common sense. He wrote: "I think very highly of common sense. In fact, I think that all philosophy must start from common sense views and from their critical examination."

But what, for Popper, was important about the common-sense view of the world was not the kind of epistemology associated with the empiricists—who thought that knowledge was simply built up of sense impressions—but its *realism*. This is the view, he wrote, "that there is a real world, with real people, animals and plants, cars and stars in it. I think that this view is true and immensely important, and I believe that no valid criticism of it had ever been proposed." (Miller 1983: 105).

Science therefore is not simply a repudiation of common-sense realism, but rather a creative (and imaginative) attempt to go beyond the world of ordinary experience, seeking to explain, as Popper put it, "the everyday world by reference to hidden worlds". In this, it is similar to both art and religion. What characterizes science is that the products of the human imagination and intuition are controlled by rational criticism and empirical validation. "Criticism curbs the imagination but does not put in in chains." (Popper 1992: 54). In epistemological terms, realism thus entails what has been described a ratio-empiricism (Bunge 1996: 322-325) or transcendental realism (Bhaskar 1975).

What exists, and how the world is constituted, depends, of course, on what particular ontology or worldview (to use Dilthey's term) is being expressed, although in terms of practical knowledge and social praxis the reality of the material world is always taken for granted, for human survival depends on acknowledging and engaging with *this* world.

It is important then to defend a realist perspective, one that Karl Marx described long ago as historical materialism. It is a metaphysic that entails a rejection of both contemplative materialism (or positivism)—the assumption that there is a direct unmediated relationship between consciousness (or language) and the material world—and the kind of social constructionism expressed by postmodernist scholars. The latter is just old-fashioned idealism wrapped up in modern guise, the emphasis being on culture, language, and discourses rather than individual perception (Berkeley) or universal cognitive structures (Kant). Such a realist ontology, while fully recognizing the significant social and cognitive activity of the human subject, also emphasizes the ontological independence and causal powers of the natural world. As Mark Johnson simply put it: "How we carve up the world will depend *both* on what is out there independent of us, and equally on the referential schemes we bring to bear, given our purposes, interests and goals." (1987: 202).

Our engagement with the world is thus always to some extent socially mediated. Equally important is the fact, as Marx put it, that we are always engaged in a "dialogue with the real world" (1975: 328). It is thus necessary to reject all forms of idealism, including Cartesian dualistic metaphysics and the kind of social constructionism expressed by postmodernist culture theorists, as well as all forms of mechanistic or reductive materialism. These include positivism and objectivism (Wittgenstein's picture theory of language), behaviourism (well-expressed by B.F. Skinner), and eliminative materialism (Churchland). But as I have explored elsewhere (2014A) many classical sociologists and anthropologists, as well as some humanistic scholars, have long attempted to steer an approach to metaphysics that avoids both the Scylla of idealism and the Charybdis of reductive materialism. This implied a dialectical approach in which thinking (culture) and being (nature) are distinct, yet at the same time "in unity with one another" (Marx 1975: 328, Sayers 1985: 15).

Again, Johnson expresses this duality rather well: "Contrary to *idealism* we do not impose arbitrary concepts and structures upon an undifferentiated, indefinitely malleable reality—we do not simply contract reality according to our subjective desires and whims. Contrary to *objectivism* (i.e. naïve realism) we are not merely mirrors of nature that determines our concepts in one and only one way." (1987: 207).

It is, however, quite misleading to interpret realism (or universal knowledge!) as if it were some ahistoric "God's eye view" of the world, or that it denies either the diversity and historicity of human social life or the plurality of perspectives on the world. Not only Nietzsche, but also people in their everyday life and the sciences generally, have long recognized what has been described as "perspectivism": the idea that our knowledge of the world is always socially mediated and that we always view the world from a particular point of view (Bunge 2003: xii). But as John Searle argued—and we quote again his thoughts—perspectivism is not inconsistent with either realism or the doctrine of epistemic objectivity that says we have direct perceptual access to the real world (1999: 21). Thus, for example, a termite may be viewed from a number of different perspectives: as an important source of food (relish), as a medicine, as a serious household and agricultural pest, and/or as having a religious or symbolic significance within a particular community (Morris 2004A).

The well-known adage of Nietzsche that "there are no facts only interpretations", endlessly quoted with approval by postmodernist scholars as if it was some profound insight, is, of course, utterly facile. It also implies that Nietzsche was a subjective idealist and relativist—although in fact (contra hermeneutic scholars) Nietzsche can be better understood as an Enlightenment thinker and philosophical naturalist. Each of the perspectives above, with relation to termites, presuppose the existence of termites as concrete living organisms, and depend on prior factual knowledge of termite biology and the local culture (Nietzsche 1956: 255; Richardson and Leiter 2001; Morris 2014A: 565-567; on scientific perspectivism see Giere 2006).

There has, of course, been a welter of debate around the issue of realism, particularly among philosophers of science. It has been stressed that the so-called anti-realists never denied the reality of the material world existing independently of humans. The debate rather was around epistemological issues, on the way in which material entities are related to human perception and cognition. But following Kant, the anti-realists (e.g., Rorty, Latour), as indicated above, misleadingly implied that the material world (nature)—and facts about this world—had no relevance or role to play in the validation of objective knowledge. Really! (On this debate, see Rorty 1982; Putnam 1990; Bunge 2006; Grayling 2007.)

There is, of course, an intrinsic relationship between realism (as defined above) and philosophical materialism, and it is to the latter doctrine that I shall now turn in the second part of this manifesto.

PART TWO

VARIETIES OF MATERIALISM

Chapter Five

Philosophical Materialism

This chapter provides an account of philosophical materialism as a worldview. After a discussion on the relationship between materialism and naturalism, I stress that a materialist ontology contends that the world consists exclusively of material things, whether physical objects, organisms (including humans), or social systems. For a materialist, such concrete entities are constantly involved in processes of change—to be material is to become—and all things are interrelated in complex ways to other material entities. In the final section I discuss, in turn, the relationship between material things and properties, relations and events, emphasizing the need for a dialectical approach to such phenomena.

Materialism and Naturalism

Philosophical materialism, as a reflective mode of thought, probably has its origins—as empirical naturalism—at the beginning of human symbolic culture around fifty thousand years ago, and perhaps even earlier (Barnard 2012: 13). It may have emerged alongside the formation of imaginative religious cosmologies. This is speculation, but certainly materialism, as a philosophical doctrine, was evident in many of the early agrarian civilizations, specifically those of China, India, and Greece.

Although scholars since the days of Rudyard Kipling continue to erect a radical dichotomy between Indian culture—with its religious mysticism and its "intense spirituality" (as Radhakrishnan [1933:24] described it)—and the supposed rationalism and materialism of Western culture—"Oh! East is east and west is west, and never the twain shall meet"—this dichotomy is completely fallacious. On two counts.

Firstly, throughout the past two millennia, materialism (and its attendant atheism) has been very much a marginal tradition within Western culture. In fact, materialism has been ignored, ridiculed, and vilified, even within the university setting, throughout much of European history. Individual scholars professing to be materialists have invariably been "demonised", excommunicated from the religious community (as with Spinoza), and denounced or harassed as heretics. For many centuries Western philosophy was simply the handmaiden of theology and supportive of the political order. As I noted earlier, all the well-known or hallowed Western philosophers—Plato, Aquinas, Descartes, Leibniz, Kant, Hegel, Wittgenstein, Husserl, Heidegger, and Whitehead (for example)—have been essentially *religious* thinkers: idealists not materialists.

Secondly, rather than Indian culture being essentially "spiritual" as Radhakrishnan (1932) and many later scholars have stressed, this perspective is something of a "myth", probably encouraged by the German romantic movement of the early nineteenth century. For as scholars such as M.N. Roy (1940), and Debiprasad Chattopadhyaya (1959) have shown, the dominant tendency of early Indian philosophy was not spiritual or mystical (as Advaita Vedanta), but rationalist and materialist—an "atheistic naturalism" (Roy 1940: 93). This materialism was exemplified by early Buddhism, Samkhya philosophy, and by the Lokayata—the "philosophy of the people"—against whom the religious clerics continually railed. Such philosophies expressed a form of materialism in being critical of religious conceptions and ritual, in defending the reality of the material world, and in emphasizing that sense impressions and a rational logic were the primary sources of knowledge, not religious faith (for the masses), or mystical intuition (for the elite) (Roy 1940: 90-93; Chattopadhyaya 1959:9; Morris 1990A).

Within early Greek culture materialism was also a significant philosophical tendency, and even Aristotle, as Roy put it, "stood with one foot in the social ground of materialism" (1940: 40). The early pre-Socratic philosophers like Heraclitus and Thales—whom Aristotle described as "students of nature" or naturalists (*physiologoi*)—the atomic theories of Leucippus and Democritus, and the philosophy of Epicurus—which is beautifully depicted by Lucretius in his poem *De Rerum Natura* (on the nature of things)—are all exemplars of a materialist metaphysic.

Since then materialism has continued to develop within Western culture, but until the Enlightenment (discussed above) it was very much a marginal philosophical tendency. I shall focus in this manifesto on three contemporary movements within philosophical materialism: the *evolutionary naturalism* of Charles Darwin, as this has been developed within Neo-Darwinian theory, which is largely a form of reductive materialism; the *dialectical materialism* that stems from Karl Marx and which emphasizes the importance of a dialectical (relational) epistemology; and the *emergent materialism* which is particularly associated with the work of Mario Bunge and other "systems" theorists.

But two initial points need to be made. The first is that I shall treat the concepts of materialism and naturalism as virtual synonyms. The term *naturalism* has two quite distinct meanings. On the one hand, it is often employed to refer to reductive materialism or physicalism: the tendency, especially among sociobiologists, behaviourists, and eliminative materialists to reduce consciousness, human social life, and culture to "nature"—to the biological or physical realms of being (Skinner 1953; Wilson 1975; Churchland 1984; Bunge 1996: 298, 2010: 96-102).

On the other hand, naturalism has been described as a worldview that affirms that the natural world constitutes the whole of reality: that there is no divine, transcendent, or supernatural realm of being, nor are there supernatural beings, whether in the form of God, geist, deities, ghosts, fairies, witches, or malevolent spirits. Mental life and consciousness, have, of course, a reality, but only in relation to the activities of humans or other forms of organic life. Spiritual beings are simply the product of the human imagination and human social life. That is to say these be-ings have no reality; nevertheless, a meaningful life for humans is possible without recourse to God or other spiritual beings (Nielsen 2001).

The Bavarian philosopher Ludwig Feuerbach expressed the essence of natur-alism succinctly: "Independently of philosophy, nature exists by itself. It is the ground on which man (humans), himself a product of nature, grows. Outside nature and man there does not exist anything. The higher beings are the creatures of our fantasy. They are merely the fantastic reflections of our own being" (quoted in Roy 1940: 54).

There is no doubt that Feuerbach's writings had a profound influence on Marx and Engels, for Engels wrote that they "placed materialism on the throne again" (Marx and Engels 1965: 592). Naturalism is therefore a philosophy of *immanence* which transfers the attribute of creativity from some spiritual being (God or spirits) to nature itself (Jonas 1966: 96).

It is, however, a common practice among anthropologists to equate natural-ism not only with ontological realism, but with Cartesian dualistic metaphysics, the reductive materialism of the sociobiologists, and Western philosophy gener-ally—the "modern ideology" (Descola 1996). This may be conducive to a struc-turalist analysis of the "exotic other", but it is quite misleading and ultimately obfuscating. Nature, of course, is not a "fetish", still less a "transcendental object" (1996: 97), but the earthly home of humans and other life forms. Needless to say (contra Descola) that both "totemism", as a symbolic schema, and "animism" (whether interpreted as a belief in spiritual beings, the endowment of animals, or other natural beings with human attributes) are both intrinsic to Western culture, indeed to all human societies. It is in this second sense that naturalism is a virtual synonym of materialism.

A second point is that philosophical materialism should not be equated with the kind of materialism that extols the acquisition of material goods, possessive individualism, or the exclusive pursuit of pleasure and profit. Philosophical mater-ialism is something very different. When he remarked on the widespread prejudice against the word materialism Engels wrote: "By the word materialism the Phil-istine understands gluttony, drunkenness, lust of the eye, lust of the flesh, arrog-ance, avarice and profiteering.... all the filthy vices in which he himself indulges in private." (Marx and Engels 1968: 600).

Indeed, it is rather ironic that philosophical materialists like Epicurus and Spinoza led rather austere lives, while a spiritual guru like the Bhagwan Ra-jneesh—a true philosophical idealist, and Hindu mystic—had promiscuous sexual encounters with hundreds of female devotees, had assets in excess of $100 million, and reputedly owned ninety-three Rolls Royces! Rajneesh described himself as a "materialist spiritualist"!! (Morris 1993; Bunge 2010: 121).

What then is philosophical materialism?

Materialism as a Worldview

Philosophical materialism, as classically defined, is the view that the world is composed entirely of matter—that the origin of everything that exists is material. Since the development of modern physics, with the focus on such entities as sub-atomic particles and the fields that hold them together, there has been a widespread belief—particularly with regards to the Copenhagen interpretation of quantum mechanics (Bohr 1958)—that the world has been "de-materialized". This idea has been eagerly embraced by subjective idealists, by fideists who wish to reaffirm the existence and relevance of some deity, and by Advaita Vedantists who, following Sankara, declare that the material world is an illusion and that reality can be identified with Brahma as pure spirit or consciousness. As a contemporary religious guru and devotee of Advaita Vedanta (Hindu esoteric mysticism) Deepak Chopra puts it: "the whole universe *is* God's mind" (2006: 220).

Ironically Sankara used concrete material objects—a rope being mistaken for a snake—to demonstrate that the material world is unreal, an illusion (*maya*)! (K. Sen 1961: 83; Morris 1994: 76-77).

But, as many scholars have argued, the religious or idealist interpretation of the findings of modern science—specifically quantum physics and relativity theory—is quite mistaken. Modern physics does *not* support the thesis that the material world has been spiritualized—or reduced to the mind, whether that of the human subject or that of God (Levy 1938: 31-32; Woods and Grant 1995: 114; Bunge 2001: 53-54, 2010: 41-42). Among contemporary materialists, "matter" is used in a more general way to refer to the fact that the real world is a system composed of material things that exist independently of human cognition.

Thus, for a philosophical materialist, the world consists exclusively of concrete or material entities—rocks, mountains, rivers, forests, living organisms with a myriad of different forms and life-ways, and, of course, human beings and their artefacts and cultures. Materialists, however, do not deny the existence or the reality of time, space, life, mind (consciousness), or culture (including ideas about gods or spirits); they simply deny that these have an autonomous existence or reality completely independent of material entities. The notion that a materialist (or socialist!) believes in the existence of a "lifeless world of matter" (McGilchrist 2010: 401) is nothing but pure religious propaganda!!

Time, for example, is not absolute. It is not a kind of container in which things are placed or events happen (Newton); nor is time subjective, simply an empirical intuition, a "form of sensibility" that precedes all our perceptions of the actual world (Kant); nor, finally, is time a thing that flows (Bergson), for only real material things, like rivers, glaciers, or volcanic lava, flow. But contrary to the theory of the Neo-Hegelian idealist John McTaggart (Whitrow 1975: 136) time is *real*. It is an ontological category that refers to the flow of events, that is, the change or becoming of real material things. Thus time, and forms of temporality, have existence and meaning only in *relation* to material things. "Time is an objective expression of the changing state of matter" (Woods and Grant 1995: 141).

Likewise, space is not absolute, but consists, as Gottfried Leibniz long ago argued, of relations between objects, including living beings, in the natural world. Of course, as a religious idealist, space only came into existence for Leibniz when God created the world (Okasha 2012: 96).

Mario Bunge therefore concluded, when he offered a materialist account of both time and space, that if there were no things there would be no space, and if nothing changed there would be no time—given that space is rooted in the separation and relations between material entities and time in the separation of events, which in themselves pre-suppose the existence of material things, whether physical objects, living beings, or social systems. Time and space, therefore, are just as material, and just as real, as the properties and relations of the material things that generate them. They do not have an independent existence, as Newton supposed. But Bunge stressed that things in themselves, and changes in their properties, also have no independent existence: "there are only mutually spaced things and successive changes in things" (2006: 245).

The real world, then, can be considered a *system* of material things, because every concrete entity interacts in complex ways with many other things, constituting what the anarchist Michael Bakunin described as a "universal causality" (Maximoff 1953: 54-56). Given that energy is a universal property of all material things, all things are changeable—everything is in flux—and all things are dynamically interrelated in various ways to many other things.

For a philosophical materialist then—to sum up—the world consists exclusively of concrete material entities, together with their properties, their relations, their actions, and their changes of state (events). All concrete entities are in a state of flux, and ever since Aristotle, all philosophical materialists have held that change (becoming) is the essence of matter. All materialists therefore have a dynamic conception of being, and articulate what Deleuze describes as a "philosophy of immanence" (Deleuze 1990). There are no inert objects, whether we consider objects in the physical world, living organisms, or social systems and institutions. All matter is in a process of change and "to be material is to become" (Bunge 2001: 50-52, 2006: 21).

I turn now to a rather abstruse but necessary discussion of the relationship between material things to their properties and their relations, and to the nature of events.

Properties, Relations, and Events

Things and Their Properties

Philosophers have long debated the relationship between, on the one hand, material things or "particulars", and, on the other hand, their properties, variously described as qualities, characteristics, attributes, or even as universals. The term "universal", however, is usually employed to describe what attributes particular material objects share or have in common.

During the medieval period, around the thirteenth century, there was a heated controversy between the Platonists, as objective idealists, who thought of universals as having a *real* existence completely independent of particulars or material things (*ante res*, before things) and the *nominalists*, such as William Ockham, who viewed universals (a property such as the colour red) as simply the names (Latin, *nomen*) that we attach to things, thereby denying their reality and emphasizing instead that only particular things existed and have reality (*post nes*, after things). Nominalism is essentially a precursor of philosophical positivism.

Earlier scholars, in particular Rene Descartes and John Locke, made the well-known distinction between primary and secondary properties (qualities) of a material thing. A primary property of a real thing exists *independently* of the human subject. Examples are the position and mass of an object, its chemical composition, the property of life, and the structure of an organism or social system. A secondary property of a real thing, in contrast, is one attributed to it on the basis of sense impressions. It thus exists only in terms of a *relationship*, namely that between a living being and the material world. Such properties, often described by philosophers as *qualia*, include feeling cold, hearing the sound of a skylark singing, the smell of mint, seeing the red colour of a robin's breast, or tasting sweet wine. All such *qualia* are subject-dependent, arising from our *relationship* with the material world as well as being derived from processors within a subject's central nervous system (Bunge 2006: 6).

But, of course, the *qualia*—our sense impressions—are not purely subjective; the song we hear is that of a skylark, the smell of mint we experience was emitted by the plant, the taste is that of wine, and the red colour we see inheres in the robin's breast (*in res*, in things). There is thus, in a sense, a dialectical relationship between concrete things and the properties we experience. For though distinct, there are no particular objects without properties, and no properties without the material things with regard to which they are intrinsically related—whatever Plato may have thought.

There has been a tendency for the natural sciences—specifically physics, chemistry, and molecular biology—to focus on primary properties, while phenomenology and radical empiricism tend to give secondary properties primacy. However, what is needed in both the biological and social sciences, including anthropology, is an approach that focusses on both the primary and secondary qualities (or aspects) of material things, whether with reference to organisms or to social and cultural systems. Such an approach should seek to develop what Brian Goodwin (1994) describes as the "science of qualities" (Bunge 1999A: 221, 2006: 13).

Significantly, this is a key distinction in a recent text advocating a speculative (i.e., philosophical) materialism, although in critiquing what he describes as "correlationism"—that is, Neo-Kantian subjective idealism and positivism—Quentin Meillassoux makes no mention at all of any materialist past or present! Instead he seems to engage entirely with the likes of Descartes, Kant, Hegel, and Heidegger—all onto-theologians, to use Heidegger's own term (Meillassoux 2008: 12-13). However, given his platonic emphasis on "hyper-chaos", "immortality", and a "future god", it is doubtful if Meillassoux can really be described as a materialist (Gratton 2014: 65-84).

When ordinary mortals—like myself!—through their common-sense understandings of the natural world, recognize a bird as a robin (an identity), and describe it as having a red breast (an attribute), or as nesting in the hedgerow (a relation)—thereby offering as a representation—they employ what Gilles Deleuze (1994) calls, rather disparagingly and with some derision, the "logic of recognition". Such a logic defines, he argues, a world of representation (to be rejected), the identity of a subject (also to be rejected), and apparently the primacy of identity over difference. Deleuze seems to view this "logic" as characteristic not only of common-sense understandings, but of Western thought from Plato to Hegel—Aristotle being described as having an "organic representation of difference" (Deleuze 1994: xv).

In his inversion of Platonism, Deleuze, like Tim Ingold (2011)—who follows in his wake—seems to feel that ordinary folk, in their experiences of the *actual* world and their common-sense understandings of this world—expressed both in their folk classifications and in a subject-predicate logic—have a fixed, sedentary, and static conception of material things. Apparently, in our common-sense understandings, and in recognizing the *identity* of material things, people view such things as static (unchanging), inert, and as self-existent, bereft of relations. Or so Deleuze seems to imply. No wonder he describes our common-sense understandings of the actual world we experience as one that is dogmatic, stupid, and illusory (Deleuze 1994: 265-269).

But, of course, when ordinary mortals identify and recognize a living thing like a robin, they are fully aware that it is a complex organism (a multiplicity), that it is continually undergoing change (becoming), and that its very survival depends on its dynamic relationships with many aspects of, or things in, the material world—food, territory, and nesting material (forming what Deleuze describes as an "assemblage"!). Indeed, in over-reacting against Platonism, Deleuze overemphasises and almost sanctifies "difference" (however defined), for difference and identity are dialectically interrelated. Difference entails the *recognition* of the identity of what Deleuze describes as a "thing", while identity always entails a recognition of *difference*. But Deleuze admitted that he always had an aversion towards things (1995: 160)—although it is difficult to know how we can avoid them in negotiating our way through life. Emphasizing morphogenetic processes and the virtual realms presupposes of course the *identity* of some individual material thing, whether physical, chemical, biological, or social.

Two points, however, may be made here. For one, even to exist is to be some*thing*, and hence to have an identity—as the objectivist Ayn Rand always stressed (Peikoff 1991: 6-7). Indeed, all living organisms strive to retain the integrity of their being and their identity (Jonas 1966: 83). Our very survival in fact depends on employing the logic of identity (recognition), for people who could not instantly recognize natural entities for what they were—whether dangerous, toxic, or edible—would simply not have survived (Margulis and Sagan 2007: 78).

The other is that ordinary folk have long recognized that organisms, people, and societies change all the time. They hardly need elitist philosophers like Deleuze, or anthropologists and literary theorists, to inform them of this obvious fact!

Things and Relations

A dialectical relationship also exists between material *things*—whether physical objects, living beings, or social systems—and their *relations*. But there has been a tendency among contemporary scholars to go to extremes.

On the one hand there are those who are described as "object-orientated" philosophers, who tend to downplay or ignore relations between phenomena. On the other hand there are the "structural realists" or "meshwork" theorists, who tend to view material entities, including human beings, as simply a "node" or an "ensemble" of relations. They thereby tend to "dematerialize" or even to virtually "eliminate" material objects (or the human subject) from their analysis, in favour of "structures" or "lines", advocating an extreme "relational" ontology (French 2014).

A noteworthy example of the first approach is that of Graham Harman (2010), who stridently affirms an "object-orientated" form of realism. His approach combines the phenomenology of Husserl and Heidegger—he interprets both as "object-orientated" philosophers—and the actor-network theory of Bruno Latour, whom he applauds as the "prince" of network theory (Latour 1993; Latour et al 2011).

Harman defines an "object" as "any thing with some sort of unitary reality". As with Latour's conception of an "actor", this definition covers not only material entities like atoms, trees, and lumps of granite, but also armies, banks, and "fictional" objects such as Donald Duck, Harry Potter, and presumably, Buffy the Vampire Slayer. Like Latour, Harman therefore *conflates* material entities, such as physical objects and social systems, with conceptual objects, such as fictional characters. The latter, however, are not actors or substances (a concept Harman is keen to reaffirm), for they possess no energy, do not change, and have no efficacy or agency apart from the actions of humans or social groups. The anthropologist Richard Shweder made a similar claim to that of Latour and Harman when he suggested that "ghosts, spirits, demons, witches, souls and other so-called religious or supernatural concepts are, in some important sense, real and objective" (Shweder 1986: 172).

Such entities, however, are only real as cultural concepts. When fervent evangelical Christians go around eradicating witches, they do not kill "real" witches, only elderly women (usually) who are completely innocent of any of the alleged crimes. To equate Donald Duck with the real, ubiquitous mallard, and to reify ideas—to impute reality to Buffy the Vampire Slayer and to other products of the human imagination—is, as Mario Bunge intimated, to engage in "magical thinking" (1996: 330).

Harman defined his own approach—object-oriented ontology or philosophy—largely in critical dialogue not only with the phenomenologists Husserl and Heidegger but also with Bruno Latour. He describes this old-fashioned positivist (as Latour described himself) as a "philosopher of relations" (Latour et al 2011: 29), suggesting that Latour tends to define objects in terms of their relations (2010: 200). In contrast, Harman advocates a more object-oriented approach, emphasizing that there must be something doing the relating, and thus that objects, not networks (relations), ought to be given primacy. He concludes that his own model "allows for entities to exist apart from all relations" (2010: 206).

But, of course, as Whitehead long ago stressed in emphasizing the "interconnectedness of real things", no material entity is self-existent. No material thing or organism could exist without relations to other things that constitute the real world we experience. Given this fact, the natural world exhibits a "togetherness of things" (Whitehead 1933: 234).

However, not only Harman but other scholars seem to advocate "object-oriented" realism, placing a crucial emphasis on material objects and their properties, but neglecting or downplaying their *relations*(Psillos 1999; S. Mumford 2012; for a critical discussion of Harman's object-orientated ontology, see Gratton 2014: 85-107). At the other extreme, however, are the proponents of what Harman describes as "naïve relationalism" (2010: 115): the tendency to reduce material objects, specifically organisms, to their relationships. To suggest that material things—including humans—are simply a "node", a "collection", or an "ensemble" of relations, whether natural or social. A noteworthy example of this extreme approach is the "meshwork" theory of Tim Ingold (2011, 2013).

Fundamentally an inverse Platonist like Deleuze, and thus prone to dualistic thinking—as I indicated long ago (1990B)—Tim Ingold tends to set up a completely false dichotomy between being and becoming, between a materialist ontology (outlined above) which highlights the reality of material objects (including organisms) and their properties (being), and what he describes as a "relational" epistemology, with its extreme focus on relations (becoming). He thus quite misleadingly *mis*interprets the materialist approach as implying that organisms are "discrete" and rigidly "bounded", bereft of relations, and that being (what something is) implies a static ontology.

Indeed, Ingold sets up a completely misleading dualism between humanity (or a robin!) as a noun (what some*thing* is—being), and as a verb (what something does—becoming), extolling the latter and denigrating the former. All this is quite facile, although it might be a useful rhetorical ploy to portray Neo-Darwinian scholars as having a static ontology. With Ingold we are therefore presented with two extremes: either the organism as an isolated, bounded entity, bereft of relations (which no materialist has ever suggested!) or—as he argues—the organism reduced to an ensemble or "knot" of abstract relations or lines, straight or wayward, with no *material* integrity or agency (Ingold 2013: 10-18).

For an emergent materialist, in contrast to Ingold, the radical dichotomy between being (what something is) and becoming (what something does) is completely redundant, as well as obfuscating. We think because we have brains; we have brains so that we can think—and so negotiate the world. For material *things*, including humans, and their *relations* (or activities) are dialectically interrelated. There are no things independent of relations (as Harman misleadingly implied), nor are there relations independent of things. As Murray Bookchin wrote—with Bateson and Capra in mind—and in emphasizing the physicality of natural things: "Abandoning the study of things—living or not—for the study of relationships between them is as one-sided and reductionist as abandoning the study of relationships for the things they inter-relate" (Bookchin 1995A: 114).

Interestingly, in contrast to Harman, Ingold interprets Latour's actor-network theory as an "object-orientated" approach that downplays the importance of relations (Ingold 2011: 89-97). But his own "meshwork" theory seems to run completely counter to Ingold's other more insightful writings, for in embracing "development systems theory" (Oyama et al 2001), he emphasizes the dynamic self-organization of organisms, including humans, and the dialectical relationship that pertains between an organism and its environment.

Setting up a radical *opposition*, as do both Deleuze and Ingold, between the *"logic of recognition"*—as reflected in the identity of material things and their properties, and as expressed in language via the subject-predicate made of expression (being)—and the *"logic of relations"*—emphasizing the primacy of relations over things (becoming)—is quite unhelpful. For both perspectives form an integral part of our encounter with the natural world, and our understanding of this world. In fact, both reflect the functions of the two hemispheres of the human brain: the left hemisphere being concerned with focussed attention, the identification of specific material entities, and language; the right hemisphere with relationships, visual imagery, and the creation of patterns. It is well-illustrated by the fact that a partridge has both to focus on finding—specifically identifying—items of food, and being also constantly aware of its surroundings—relations—vis a vis material objects and other living organisms, specifically predators. These two modes of relating to the world are dialectically interrelated, and in our everyday experiences of the world, both hemispheres are continually involved. It is debatable whether our identification of specific material things or living creatures, or our use of language to describe things and events, implies that humans ever experience the world as fixed, inert, and lifeless—the model characteristics of the left hemisphere, according to McGilchrist! (McGilchrist 2010: 31).

Critiquing the subject-predicate mode of expression as "one-sided" because it does not refer to relationships (Bateson 1979: 72-73) seems to be equally facile, given that we always tend to talk to other humans. If we talked to woodpigeons we would perhaps not describe ivy berries as bitter and nauseous! Relationships are not prior, nor do they "precede" material entities and their properties (Bateson

1979: 46-47). Ivy berries are unappetizing for humans not because of some abstract "relationship", but because of the material composition, internal relations, and dynamics of both the ivy plant and humans. Significantly Bateson always tends to put things in inverted commas—"things"—as if they were either simply human constructs, or constituted solely of relationships. No wonder Bookchin critiqued Bateson's systems theory as "dissolving" the *materiality* of life-forms into relationships, that were both abstract and lifeless" (1995A: 115).

Things and Events

The relationship between concrete material *things*, whether physical objects, living beings (including humans), social systems, or *events*, is one that is complex. It has long been a tropic of abstruse debate among analytical philosophers (Davidson 1970; Armstrong 1997). From a materialist perspective—one that is dialectical—things and events can neither be conflated, nor treated as if they were opposed concepts that belong to different realms of being.

The process philosopher Alfred North Whitehead tended at times to describe material things, or actual entities and events (actual occasions), as if they were virtually synonymous, though his main emphasis was giving primacy to events. He thus suggested that material things, or what he called "societies" to indicate their systematic character—enduring physical objects, living organisms, and human persons are specifically mentioned—were simply a "nexus" of events. Such things had a unity and a reality, but only as a nexus or "set of occasions". An actual occasion (event), for Whitehead, was seemingly ephemeral and had no history; it only "becomes and perishes". In contrast a "society"—an object, organism, or human person—enjoys a history, one that is expressed in its reaction to changing circumstances. In emphasizing that the world is composed of events (actual occasions) Whitehead tends therefore to "de-materialize" the concrete material entities that constitute the material world (Whitehead 1933: 199-205; Bunge 2001: 54).

A similar perspective was expressed by the early Wittgenstein who asserted that the world was not composed of material things—such as stones, teacups, subatomic particles, organisms, and people—but was a "totality of facts". A fact, of course, Wittgenstein defined as a "state of affairs"—that is, an event—which *pre-supposes* the existence of concrete material entities (Wittgenstein 1922; Grayling 2001: 40-42).

Often linked with Whitehead and Wittgenstein, John Dewey is sometimes depicted as an idealist philosopher who downplayed the salience of the material world. But as I have argued elsewhere, as a Darwinian scholar and as an advocate of empirical naturalism, Dewey always envisaged a dialectical relationship between material things and events, while he affirmed the eventful nature of all existence. In fact, in his critique of Bergson's vitalism, Dewey stressed that "what exists are things acting and changing". Thus for Dewey, all things have a history, and all things are entities in process (Dewey 1934: 214; Novack 1975: 50-60; Morris 2014A: 148-154).

The enigmatic Gilles Deleuze once remarked: "I have it's true spent a lot of time writing about this notion of event; you see, I don't believe in things." (Deleuze 1995: 160). It is therefore quite common for his literary devotees to declare that the world for Deleuze does not consist of material things, but events. Thus Deleuze's transcendental empiricism aligns with Whitehead's process philosophy.

Yet the same Deleuze also stridently affirmed (with Felix Guattari) that matter on the earth was a "*body* without organs" and that everything in the world was a "machine". Thus we have, in their obscure and exuberant terminology—energy-machines, organ-machines, technical machines, desiring-machines (humans!), and socio-machines (social systems). Everything in the world, including signs, human emotions, and the products of the human imagination—like Buffy the Vampire Slayer!—are, according to Deleuze's worldview, machines, linked together by mechanic connections in a universal causality. For every machine affects, and is affected by, other machines, such that the whole world of nature is a "process of production". Interestingly events as such hardly get a mention in *Anti-Oedipus*. The book is, in fact, a strident illustration of reductive mechanic materialism (Deleuze and Guattari 1977).

Yet in Deleuze's key texts outlining his own transcendental empiricism, particularly *The Logic of Sense*, there is hardly any reference to machines or mechanic connections, only to bodies and events. In fact, in discussing Stoic philosophy, events take centre stage. It therefore has to be recognized, as Slavoj Zizek emphasized, that Deleuze's ontology consists of two contrasting perspectives or "logics" that seem to co-exist in his work (2004: 21).

On the one hand here is the *logic of production* that involves the becoming of actual *things*, described by Deleuze as "assemblages". This involves morphogenetic processes of "individuation" or "actualization", and the production of individual beings—whether objects, organisms, human subjects, or social systems. According to this logic of production, events or intensive morphogenetic processes take place in a virtual realm that is "incarnated" within a particular thing. The actualization occurs, Deleuze writes, "in things themselves" (1994: 214-215). Becoming, for Deleuze, relates to "spatio-temporal dynamics" at a virtual (molecular) level within the material thing, and this, he felt, was quite distinct from its history, which he continually abjured. We are informed that the virtual realm, the plane of immanence, is "populated only by events", not apparently by elements—constituents—and their relations (structure).

On the other hand, there is a *logic of sense*, which involves the becoming of *events*, which are the *effect* of bodily-material processes. By events, or "singularities", Deleuze indicates a motley of diverse items: for example, bottlenecks, knots, point of fusion, condensation, sickness, joy, health, hope, and anxiety (2004: 63). Elsewhere an event is described as a "haecceity" (something given)—"a draft, a wind, a day, a time of day, a stream, a battle (or) a cloud of locusts carried by the wind" (Deleuze and Parney 1987: 69; Deleuze and Guattari 1988: 262; Marks 1998: 39).

Such events or singularities are described as incarnated in states of affairs, and deemed to be ideal, incorporeal, immaterial, mobile, impersonal, and as happening only on the "surface"—specifically of an organism (Deleuze and Guattari 1994: 156; Deleuze 2004: 119, 160).

Whereas the individual thing, for Deleuze, is fixed, and sedentary (2004: 122), events as singularities are nomadic, prior to the individual or person, and are involved in their genesis. Therefore, as far as Deleuze is concerned, a flesh wound or a smile produces the individual. The human person does not smile: the smile generates the person—or so Deleuze, following Nietzsche would have us believe (Deleuze 2001: 6, 2004: 118-122). Heralded as a reversal of Platonism, the logic of sense seems to verge on idealism.

Events—like properties, relations, and ideas—do not float around as immaterial entities completely independent of material things, as Deleuze seems to imply. Even more mystifying is his notion that events are "imprisoned" in actual things (Hallward 2006: 45). A smile has no meaning or reality apart from the person who smiles, and the social group to which they belong. Likewise, a specific armed conflict or rebellion has no meaning apart from its earthly location, the people participating in the conflict, the weapons used, and the social groups (systems) involved in what constitutes the *event*—the battle. To suggest that the battle is an "incorporeal event" completely independent of human beings—fixed or otherwise—and the material world, may be going beyond common-sense understandings but it can only be described as obfuscating.

The classic example of reducing material entities to their actions (events) relates of course to Nietzsche. Observing that ordinary mortals, whom he denigrated and despised, often use such phrases as the "lightning flashes", Nietzsche in his musings suggested that there is no material substratum behind the action or event. He thus writes: "there is no "being" behind doing, effecting, becoming: the "doer" is merely a fiction added to the deed—the deed is everything." (1956: 178-179).

This is problematic, to say the least, for it virtually denies the reality of people—who do the imagining!—living organisms, and material objects such as comets and atoms, all of which have the power, by virtue of their various internal structures, to bring about various kinds of actions (events). As the critical realist Andrew Collier wrote: "No event or action exists before it occurs, but its agent always does. A battle does not first exist and then is fought, but the armies do first exist and then fight" (1994: 9).

Likewise, thinking is not the brain. Thinking is one of the functions of the brain; it is what the brain does! Nietzsche and his postmodern acolytes like Judith Butler have things back to front. Of course, Nietzsche rejects such subjective idealism in other contexts, and he has been interpreted as fundamentally a philosophical naturalist and as "deeply and pervasively Darwinian" (Richardson 2004: 14; Morris 2014A: 575).

An event (or a process) (contra Nietzsche, Whitehead, or Deleuze) is something that happens, an occurrence, a history, a state of affairs, and it always *pre-supposes* the existence of a material thing(s). Events may, of course, be regarded as "quasi-causes" (as Deleuze describes them); that is, giving rise to other events—this is the classical definition of causality (Bunge 1996: 31)—so that, for example, a crop failure may lead to famine. But this hardly suggests that events are always contingent, as Deleuze seemed to imply. Indeed, an emphasis on the contingency of events explains nothing, and hardly helps our understanding of such events as famines or political rebellions against the colonial state, for example (Morris 2016).

An event, like the proverbial cat sitting on the mat, *pre-supposes* the existence of at least three concrete material things, namely the cat, the mat, and the ground beneath the mat (what Deleuze would describe as an "assemblage"). An event is thus a change in the state of some material object. There are no events in themselves, and all events (or processes) pre-suppose the existence of material things. Events and concrete material things are thus dialectically interrelated. There are no immaterial events; nor are there unchanging, fixed material things, actual or otherwise (Bunge 2001: 54, 2006: 16).

I shall now critically outline, in turn, each of the three contemporary forms of philosophical materialism, beginning in the next chapter with Darwin's evolutionary naturalism.

Chapter Six

Darwin's Evolutionary Materialism

The year 2009 marked the two hundredth anniversary of the birth of Charles Darwin (1809-1882). It was celebrated by an absolute welter of books on every aspect of his life, as well as on the theory of evolution and the place of Darwin in the history of science and Western thought more generally. Along with Marx and Freud, Darwin has indeed had a profound influence on the zeitgeist of the twentieth century, his theory of evolution even being described as the "creation myth" of our own culture and era. It is suggested that it provides the kind of cosmic mythology that gives meaning and structure to the modern scientific worldview (Midgley 1985: 30). Even so, Darwin's theory of evolution, which expresses a form of evolutionary naturalism, continues to be challenged and berated by Christian and Islamic theologians and by religious fundamentalists, and a veritable "anti-evolutionary crusade" still has wide currency and political support (Foster et al 2008; Kitcher 2009).

Nevertheless, towards the end of the last century Darwin's evolutionary theory was given a new lease of life, particularly after the publication of two key texts: Edward Wilson's *Sociobiology: The New Synthesis* (1975), and Richard Dawkins' *The Selfish Gene* (1976). Embraced with fervour and stridency by many scholars, including many anthropologists and psychologists, Darwin's seminal ideas thus came to be widely employed to advance the thesis that there is a biological basis to all forms of social life and culture. This approach has generally been described as ultra- or Neo-Darwinism.

After briefly outlining Darwin's evolutionary paradigm and his new conception of science, I shall focus in this chapter on sociobiology as a strident expression of evolutionary naturalism. In particular, I shall offer some critical reflections on three strands or varieties of sociobiology: evolutionary psychology, the theory of memetics, and gene-culture evolution.

Darwin's Evolutionary Paradigm

The publication of Darwin's *On the Origin of Species by Means of Natural Selection* in 1859 completely revolutionized our way of understanding the natural world. As a form of philosophical materialism that also stressed (like Marx) the importance of historical understanding, the book represented, as the distinguished biologist Ernst Mayr suggested, perhaps "the greatest intellectual revolution" experienced by humankind (2002: 9). Darwin's evolutionary theory, as Mayr was always keen to stress, combined essentially five distinct theories. Although logically distinct, these five theories were closely interrelated and formed a unity or paradigm. Following Mayr (2004: 97-115) we may outline each of these theories in turn.

Evolution as a Process

This is the idea, now widely accepted, that the earth has a long history and is constantly changing. Even in the early nineteenth century, geologists like Charles Lyell were fully aware of the great age of the earth; what Darwin and other biologists argued was that organisms were also transformed in time. The earth and its organic life were not of recent origin; nor divinely created; nor were they unchanging, or perpetually re-cycling; they were, rather, undergoing constant change. Evolution was thus a historical process, and the evidence for organic evolution—derived from the fossil record, embryology, comparative morphology, and molecular science—was considered by Mayr to be overwhelming (Mayr 2002: 12-39).

The Theory of Common Descent

This theory suggests that all organisms are descended from a common ancestor, and that all modes of life, whether animals, plants, fungi, protista, or bacteria, can ultimately be traced back to a single origin of life on Earth some 3,500 million years ago. It is important to stress that Darwin's theory was not a "historicized version" of the eighteenth century "great chain of being"—as Noske (1997: 62-63) misleadingly suggests—which even Lamarck tended to follow. This is because descent from a common ancestor, for Darwin, was a *branching* phenomenon, and entailed a multi-lineal form of evolution. As Darwin recognized, this meant that the classification of organisms into species, genera, families, and orders was not some arbitrary scheme based simply on resemblance, but rather reflected genealogical relations, or what he described as "chains of affinities". The theory of common descent thus involves both a historical dimension—genealogical connections and transformations through time—and a geographical dimension—the diversification of species in space (Darwin 1951: 474; Lovejoy 1936; Mayr 2002: 5-7).

The Theory of Gradualism

According to this theory, evolutionary change is a gradual process, in that new species evolve gradually from pre-existing forms. There is no sudden creation of a new species. As Darwin put it, the geological succession of organic forms involves "their slow and gradual modification, through variation and natural selection" (1951: 379). But important research by Niles Eldredge and Stephen Jay Gould (1972) on the theory of "punctuated equilibrium" suggests that evolutionary change may at times be rapid, involving short bouts of speciation, and that some species may exist for many millions of years without undergoing any appreciable change. This implies, of course, that species are real entities, like individual organisms; they have their birth (through speciation), lifespan (of varying periods), and their eventual demise (when they become extinct) (Hull 1978; Gould 1983: 149-154; Delanda 2006: 48-49).

The Diversification of Species

This theory deals with the multiplication of species, and thus the origins of organic diversity or what is generally described as speciation. There are around thirty million species of what Darwin described as "organic beings" (Margulis and Sagan 2007: 80). Such species, he wrote, are "tolerably well-defined as organic life is by no means entirely chaotic". Or, as Mayr put it, there are real discontinuities in organic nature "which delimit entities that are designated species." The species is therefore the basic concept of biology. The multiplication of species—as inter-breeding natural populations—is a complex issue which Darwin struggled to solve, but it essentially implies that the development of new species occurs through either geographical isolation, allopatric speciation (as Darwin recognized with the birds of the Galapagos Islands), or sympatric speciation within a population (Gould 1983: 170-176; Mayr 1991: 26-31).

Natural Selection

The cornerstone and most original element of Darwin's evolutionary paradigm was the theory of natural selection. This theory is perhaps best expressed in Darwin's own words: "As many more individuals of each species are born than can possibly survive and as, consequently, there is a frequently recurring struggle for existence, it follows that any being, if it should vary however slightly in any manner profitable to itself, under the complex and varying conditions of life, will have a better chance of surviving, and thus be *naturally selected.*" (1951: 4). Darwin suggested that Herbert Spencer's expression "survival of the fittest" was an appropriate description of the idea of natural selection (1951: 63-65).

Natural selection is thus a two-step process. The first step is that of genetic variation, the production of varied characteristics among the unique individuals of a particular species; the second is the struggle for existence, and the survival of those individuals who are best able to cope with the challenges of the environment (adapt) and thus reproduce successfully.

It is important to note that there is no progressive tendency or cosmic teleology in Darwin's conception of evolution, for it suggests a phylogenetic rather than an orthogenetic process, a process of increased organic diversity and complexity. Or, as Alfred North Whitehead famously described evolution, it is a "creative advance of nature.... into novelty" (1920: 178).

There has been a protracted debate among biologists regarding what is the unit of selection. Darwin, and many generations of biologists, have tended to consider the individual organism—the phenotype—as the fundamental unit of selection, rather than the gene (which is simply replicated). The Neo-Darwinists, however, as we explore below, with their "gene-centred" conception of biology, tend to regard the gene, not the organism, as fundamental.

As many scholars have long recognized, Darwin's theory of biological evolution, which was both materialist and historical, completely transformed our understanding of the natural world and human social life. I have discussed elsewhere the impact of Darwinian theory on our understanding of the human condition (2014A: 49-62), but a few brief points may be made here.

Firstly, Darwin's theory completely undermined Cartesian philosophy, both its dualistic metaphysics and its mechanistic conception of nature. For in its suggestion that humans originated in Africa and were descended from ape-like ancestors, Darwin emphasized the fundamental continuity between the natural world and humanity. In fact, contemporary studies have confirmed that humans and chimpanzees shared a common ancestor in Africa around five million years ago, and that it is estimated that we humans share 98.5 percent of our DNA with chimpanzees. Humans have indeed been described as the "third chimpanzee" (Diamond 1991). Expressing a thorough-going materialism, Darwin also undermined the mind/body dualism that was also intrinsic to Cartesian philosophy, suggesting—in contrast to Alfred Wallace's spiritualism—that all mental activity is intimately connected with the functioning of the human brain (Gruber 1974: 218).

Equally important in undermining the Cartesian mechanistic worldview and in affirming that organic evolution was both materialist and historical, Darwin placed a crucial emphasis on the importance of openness, chance, probability, and the individuality and agency of all organisms in the evolutionary process. Long before quantum physics, systems theory (Capra 1997), eco-feminism (Plumwood 1993), deep ecology (Devall and Sessions 1985), and post-humanism (Braidotti 2013), Darwin initiated a new ecological worldview, emphasizing the continuity between humans and nature.

Secondly, in introducing the notion that humans are not the special products of God's creation but evolved according to principles that operate throughout the living world, Darwin not only stressed the intrinsic organic (not spiritual) links between humans and the rest of nature, but also initiated a critique of anthropocentrism. This is the idea that humans are either God's chosen elect, given dominion over the earth and all its creatures, or the apotheosis or pinnacle of biological evolution. Either way, human beings are viewed not only as radically distinct but also as pre-ordained to possess and rule the earth, and to use it as a resource solely for human benefit. Darwin offered a critique of such anthropocentrism, emphasizing that humans are the product of organic evolution and are closely allied to other animals, specifically primates. Darwin's evolutionary perspective is therefore, as Stephen Jay Gould suggested, an important antidote to the cosmic arrogance that has long been a part of Western culture (1980: 14).

Finally, Darwin's evolutionary biology, as we have indicated, emphasized the uniqueness of the individual organism, and the important role that chance, diversity, and novelty played in the evolutionary process. But as with geology and palaeontology, Darwin's biology also completely undermined the radical dichotomy that is often made between science and history. Long before the pretentious Heidegger, Darwin emphasized the importance of time in our understanding of the world. Darwin was a historicist and a materialist. As Stephen Jay Gould has again suggested, although Darwin as a naturalist was always fascinated with the "smallest" of creatures—coral reefs, earthworms, insects, and barnacles—his greatest achievement was possibly when he outlined the principles of a *historical* science. Such a science was distinct from Cartesian mechanistic (synchronic) science, with its emphasis on experiment, prediction, and determinism. In their advocacy of both philosophical materialism and historical science, Darwin and Marx were thus kindred spirits (Gould 1984: 120-133; Mayr 1997: 113-115; Foster 2000).

It must be recognized, of course, that Darwin's historical naturalism and his theory of biological evolution were by no means unique to Darwin. Nor must Darwin be depicted as a "lone iconoclast" or as a "one-man band". For as I discussed in chapter three, many scholars during the eighteenth-century Enlightenment—especially in France—pioneered evolutionary naturalism as a worldview, introducing such conceptions as deep time, the impact of the environment on organisms, the unity and diversity of nature, and the evolution or transformation of biological species through historical time. Darwin essentially gave us a seminal and enduring expression of such evolutionary naturalism, particularly in highlighting natural selection as a key mechanism in the transformation of species. Indeed, the disciplines of geology, palaeontology, biology, and anthropology—on which Darwin drew for insights—were all rooted in the Enlightenment (Elsdon-Baker 2009: 47-72).

Sociobiology: A Gene-Centred Perspective

The Neo-Darwinian paradigm, which came into prominence towards the end of the last century, is particularly associated with the writings of Edward Wilson (1975) and Richard Dawkins (1976). It has become the focus of much debate and controversy, both for and against. In many ways, the controversy initially revolved around the ultra-Darwinism of Wilson's sociobiology—and its offshoots, evolutionary psychology and memetics—and the critiques of the Marxist-inspired biologists Stephen Jay Gould and Richard Lewontin. These two scholars, whom I shall discuss in the next chapter, aimed to develop and promote a more dialectical biology, in opposition to the ultra-Darwinism of Wilson and Dawkins. One scholar has recently declared that "Neo-Darwinism is dead" (Ingold 2013: 1). This is rather wishful thinking, for the Neo-Darwinism sect—or "thought collective", as Lynn Margulis disparagingly described it (Margulis and Sagan 1997: 28)—still flourishes in academia, and many of its insights have become a part of the popular culture. Indeed, several contemporary scholars have utilized certain Neo-Darwinian insights to enhance our understanding of human social evolution rather than getting bogged down in Nietzschean "becoming" (see, for example, Dunbar 2004; Runcinan 2009; Barnard 2016).

Sociobiology suddenly burst upon the intellectual scene in the spring of 1975, amid a fanfare of publicity. It marked the publication of *Sociobiology: The New Synthesis* by the Harvard biologist Edward Wilson. Anticipating the book would be a controversial one, its publisher gave it maximum publicity.

The book did indeed become the subject of intense debate, and Wilson suddenly found himself a celebrity—unfairly one much maligned. What caused the furore was that Wilson applied his theory of sociobiology, defined as "the systematic study of the biological basis of all social behaviour" (1978: xvi), not only to the social life of animals—from the invertebrates to the primates—but also to humans. Wilson argued that biological principles could be applied to all forms of human social life and offered biological explanations of such socio-cultural manifestations as the gender division of labour, warfare, ethics, religion, competition, altruism, tribalism, and genocide. He even described "racism" as being akin to the springtime singing of birds (1978: 70).

Although Wilson's sociobiology has been described as an intellectual "revolution" (Laland and Brown 2002: 67), it is clear that Wilson did not seek the integration of biology with the social sciences and humanities, while respecting the autonomy of human social life but rather uncompromisingly applying evolutionary biology, narrowly conceived as a "gene-centred perspective", to all aspects of human existence. As many scholars have indicated, Wilson, although a philosophical materialist, essentially proposed a reductive biological approach to human social life and culture. Indeed, he specifically described his own approach as one of "genetic determinism" (1994: 332). Wilson, of course, recognized the importance of human symbolic culture, but he nevertheless always stressed that culture is "rooted" in biology, and as he famously put it, "the genes hold culture on a leash" (1978: 167, 1997: 126).

There are therefore close affinities between Wilson's sociobiology and the evolutionary biology of Richard Dawkins, who introduced the notion of the "selfish gene". Whereas, for Dawkins, genes are "immortals" and "replicators", human beings are merely their "receptacles"—simply "survival machines" that are "programmed to do whatever is best for its genes as a whole" (1976: 66).

Oddly enough, at the close of his widely acclaimed book *The Selfish Gene*—which, like Wilson's monograph, is devoted entirely to promoting a "gene-centred" interpretation of human behaviour—we learn that the sovereignty of the selfish genes can be annulled by the human species. We—alone on Earth—have the power, Dawkins tells us, to "rebel against the tyranny of the selfish replicators"—the genes! (1976: 201; for critiques of Dawkins' selfish gene theory see Gould 1983: 72-78; Bookchin 1995B: 39-41; Woods and Grant 1995: 338-344; Elsdon-Baker 2009: 125-142; c.f. Dawkins 1976: 271-272).

Although often associated with Wilson, Dawkins denied that he was a "genetic determinist". However, like the early population geneticists, he defined evolution by natural selection as the "differential survival of genes" (1982: 18). Nevertheless, Dawkins admitted that the genes—the replicators—are not selected directly but by proxy, as they are only "selected by virtue of their phenotype effects" (1982: 117; Morris 2014A: 95).

From its inception, Wilson's sociobiology as a research strategy was subject to a welter of criticisms. Such criticisms came from scholars of many different intellectual traditions: radical biologists, cultural materialists, moral philosophers, social anthropologists, and philosophers of science (see, for example, Sahlins 1977; Bock 1980; Harris 1980: 119-140; Kitcher 1985; Midgley 1985). It is beyond the scope of this manifesto to faithfully review this literature, but two points may be made.

Firstly, Wilson tends to describe other cultures, particularly hunter-gatherers, in terms that derive essentially from his own culture. Indeed, his whole discourse is permeated with the values and ideas which C.B. MacPherson (1962) described as possessive individualism. Human beings are thus seen universally as being territorial and xenophobic, and as being aggressive, self-aggrandising, selfish creatures who are essentially concerned with maximizing their own reproductive (genetic) fitness. Even the genes they possess are described as capital, and all cooperative aspects of human life are viewed as really being a form of selfishness (Rose et al 1984: 245). As with Hobbes, Wilson seems to equate the state of human nature with the ideology of capitalism. Indeed, in a more recent study, the Neo-Darwinian scholar Matt Ridley (2010) has argued that free-market capitalism is essentially a manifestation or expression of basic human nature—as depicted, of course, by "gene-centred" Neo-Darwinian scholars. Small wonder that another Neo-Darwinian scholar, the philosopher Daniel Dennett, has described Thomas Hobbes as the "first sociobiologist" (1995: 453). Hobbes, of course, was a mechanistic materialist.

Secondly, many scholars have been critical of the genetic determinism that is implied and stressed in Wilson's brand of sociobiology, given its "gene-centred perspective". Such biological (or genetic) determinism is, in fact, specifically related to the reductive and atomistic epistemology that Wilson explicitly advocates. For Wilson views a reductionist strategy as the accepted and traditional mode of scientific analysis (1978: 13). He acknowledges the importance of environmental factors and human symbolic culture in understanding human life—besides "biological predispositions"—as well as the existence of "novel emergent phenomena", but the thrust of his writings is to reduce human behaviour and human culture to the "laws of biology".

In his later study, *Consilience* (1998), Wilson continued to reaffirm biological reductionism as a research strategy. Although the book offers a salutary defence of the Enlightenment tradition and evolutionary naturalism, it is clear that Wilson interprets consilience, the unity of knowledge, as implying "consilient with the natural sciences". He even suggests, when discussing levels of organization, that all "laws and principles" can eventually be reduced to physics (1998: 58-59). At heart, Wilson is a reductive materialist or physicalist.

The "vertical" nature of Wilson's approach—involving a chain of causation from genes to culture via epigenetic rules—is a limiting perspective, for it ignores what Owen Flanagan calls the "horizontal" dimension of human life: the cultural patterns and behaviour that are embedded in complex social relations and historical contexts. Human agency, human history, and social institutions, particularly those relative to production and political life, are completely bypassed by Wilson's sociobiology, quite apart from the complex ecological relations that humans have with nature. As Mario Bunge expressed it: obsessed with genes, sex, and reproduction, Wilson and other sociobiologists overlook the fact that while these are undeniably important, human beings are also concerned with the production of food, shelter, and politics, not simply with spreading genes (Bunge 1998: 35; Flanagan 1984: 262-266). Thus sociobiology is essentially a form of reductive materialism.

A response to Wilson's genetic determinism with its reductive tendency should not, however, imply an equally one-sided cultural determinism—as Sahlins (1977) appears to suggest (see Kuper 1999: 197-199)—and thus a denial of biology. What is needed is an integrated approach, focussed on the complex and dialectical inter-relationship between the biological and socio-cultural aspects of human life, without collapsing the distinction between biology and human social life with some amorphous "foam", "network", or "meshwork". Nothing is to be gained by *conflating* the organism (including humans) with its environment, both natural and social. This integrated *dialectical* approach is well summed up by the phrase: "Humanity cannot be cut adrift from its own biology, but neither is it enchained by it." (Rose et al 1984: 10).

Varieties of Sociobiology

In spite of the welter of criticism that surrounds Wilson's sociobiology and his Neo-Darwinian approach to the understanding of human social life and culture, it seems to have flourished in the final decades of the twentieth century. It was in fact taken up with enthusiasm by many anthropologists and psychologists, and consequently several distinct strands or approaches tended to emerge. All to some degree reflect aspects of sociobiology, testifying to the richness and pluralistic nature of Wilson's theory (Laland and Brown 2002: 106-108). I shall here offer some critical reflections on three distinct approaches: evolutionary psychology, the theory of memetics, and gene-culture co-evolution.

Evolutionary Psychology

It was during the 1990s that evolutionary psychology suddenly blossomed as an academic discipline. The edited volume *The Adapted Mind* (1992) was something of a manifesto of this new movement. The opening chapter by the anthropologist John Tooby and the psychologist Leda Cosmides, entitled "The Psychological Foundations of Culture" (1992: 19-36), was certainly a landmark text. As a research programme, evolutionary psychology was embraced by scholars from several disciplines. They seem to have formed a mutually supporting intellectual coterie. Among the key figures were Steven Pinker, Dan Sperber, Jerome Barkow, David Buss, Helena Cronin, and Donald Symonds, as well as Tooby and Cosmides. Although it is suggested that the evolutionary psychologists owed little to Wilson's sociobiology, his genetic hypothesis in fact postulated the existence of "epigenetic rules" (or mental predispositions) that mediated between genes and culture (1978: 32). The popularity of evolutionary psychology, both within the university and among the general public—given its emphasis on sex and reproduction—gave rise to a number of useful introductory texts (eg Pinker 1997; Plotkin 1997; Buss 1999; Badcock 2000; Dunbar et al 2007).

Evolutionary psychology is characterized essentially by three fundamental and interrelated tenets or principles.

The first is the suggestion that human beings *do* have a nature that defines them as a unique species, and that all psychological or social theories inevitably imply, implicitly or explicitly, a specific conception of human nature (Buss 1999: 47). Evolutionary psychologists therefore contend that anthropology, and the social sciences more generally, either misleadingly *deny* that there is such a thing as a universal human nature, or conceive of the mind as essentially a *tabula rasa*, a blank slate. Tooby and Cosmides describe this conception of human nature as the "standard social science model". It was well expressed, they suggest, in the writings of Emile Durkheim (1895), who advocated a form of sociological holism, as well as in the cultural determinism of such anthropologists as Franz Boas (1940), Alfred Kroeber (1917), and Clifford Geertz (1973). Such anthropologists emphasized the diversity and crucial significance of human cultures, but although acknowledging the "psychic unity" of humankind they invariably—according to the evolutionary psychologists—viewed human nature as an "empty vessel". This depiction of anthropologists was stridently expressed in Steven Pinker's polemical study *The Blank Slate* (2002). The Neo-Darwinist concluded that all social scientists are either naïve empiricists, deny that there is such a thing as human nature, or have a "blank slate" theory of the mind.

The second tenet of evolutionary psychology is the contention that the human mind is not a blank slate, but consists of a number of information-processing mechanisms: psychological mechanisms or predispositions that have evolved through natural selection during the course of human evolutionary history. Taking a cue from Noam Chomsky's (1975) well-known theory of an innate language acquisition device, evolutionary psychologists postulate that these psychological mechanisms underpin human social life and behaviour. Each mechanism, as Donald Symons put it, "was designed by natural selection in past environments to promote the survival of genes that directed its construction by serving some specific function" (1992: 138).

Each psychological mechanism is thus an "evolved mental mechanism" or a specific "form of adaption". These mechanisms, variously described as "mental modules" or "psychological adaptations", are conceived as both innate (unconscious) and domain-specific. It is unclear how many of these adaptive mechanisms constitute the human brain/mind, but among those viewed as significant—each relating to a specific function—are the following: mating strategies, kin recognition, maternal attachments and the ability to empathize with the thoughts and feelings of other humans, the theory of mind, language acquisition, the categorization of living forms' emotional expressions, food preferences, incest avoidance, and mechanisms related to sense perception (Tooby and Cosmides 1992: 121, 2006: 181).

Unlike other sociobiologists, evolutionary psychologists tend to downplay the importance of genes. They were fond of describing the "adapted mind" as being like a Swiss army knife, designed to solve a variety of existential problems—the problems of survival of early humans against the backdrop of the "hostile forces of nature", as Buss describes it (1999: 67; on the modularity of the human mind see Fodor 1983; Hirschfield and Gelman 1994).

The third tenet of evolutionary psychology is that the above psychological mechanisms essentially reflect adaptations, through natural selection, to an earlier hunter-gatherer mode of existence. As Steven Pinker succinctly expressed it: "The mind is a system of organs of computation, designed by natural selection to solve the kinds of problems our ancestors faced in their foraging way of life." (1997: 21).

Research studies of evolutionary psychologists have focussed essentially on psychological issues and their relationship to evolutionary biology. They embrace such topics as food preferences, mating strategies, conflict between the sexes, problems of homicide, the origins of language, and the nature of religious beliefs. Such research has emphasized the universality of much human behaviour and has tended to interpret such behavioural problems in terms of reproductive success (fitness); that is, as genetically based evolutionary adaptations. Much of this research has been highly controversial (e.g., Buss 1994; Daly and Wilson 1998; Thornhill and Palmer 2000 on, respectively, sex, homicide, and rape).

As with sociobiology, a welter of criticisms has been expressed regarding the basic premises of evolutionary psychology. I have elsewhere critically reviewed some of this extensive literature (2014A: 90-94), to which the reader is referred. But I will make two brief points here with reference to this critical literature.

The first major criticism of evolutionary psychology, and the cognitive sciences more generally, is that it tended to ignore the emotions. The Marxist scholar and neuroscientist Steven Rose, for example, has emphasized that human actions and behaviour are not simply about the brain and cognition, but also involve the body and emotions. He argues that emotions and feelings are implicated in all learning, as well as in all human interactions with the natural world. There is thus an intimate connection between human consciousness and the emotions. Indeed, Christopher Badcock has drawn attention to the fact that while Darwin was very much interested in the emotions, and wrote a pioneering study on the subject, emotions have been singularly ignored by evolutionary psychologists (Badcock 2000: 23; Damasio 2006; Rose 2005: 59-61).

Secondly, the notion of the "architecture of the mind" (Tooby and Cosmides 2006: 175) implies a rather static structure. Nothing could be more inappropriate, Rose suggests, as a way of describing the "fluid dynamic processes whereby our mind/brain develop and create order out of the blooming buzzing confusion of the world". Thus Rose concludes, in contrast to the evolutionary psychologists, that what evolutionary theory has taught us is that evolution has produced human organisms with "highly plastic, adaptable, conscious brains/minds and ways of living" (Rose 2005: 103-105). This, however, does not imply a "blank slate" theory of the mind. Indeed, given the diversity within the social sciences, Tooby, Cosmides, and Pinker's depictions of the social scientific tradition verge on caricature. It is worth recollection, for example, that Erich Fromm, who stressed a dialectical relationship between biology and social life, emphasized that human nature is "*nota* blank sheet of paper on which culture can write its script" (1949: 23). The notion that all social scientists and anthropologists are metaphysical dualists and cultural determinists (they can hardly be both!) is thus completely misplaced (Bloch 2005: 90).

Other biologists have reaffirmed the theories of the early Marxist scholars, and emphasised the extraordinary behavioural plasticity of humans (and their brains), which is enhanced by systems of symbolic communication. Thus a crucial aspect of cultural evolution is the extremely variable ecological and social environments that humans have created over several millennia. Evolutionary psychologists, however, tend to ignore the complex socio-historical and changing cultural life of humans (Jablonka and Lamb 2014: 209; for further critical discussions of evolutionary psychology see Rose and Rose 2000; Whitehouse 2001; Buller 2006; Bunge 2010: 250-253).

The Theory of Memetics

In the final chapter of *The Selfish Gene*, Dawkins, almost as an aside, turned his attention away from genes and to the importance of human culture. He recognized that culture—or learned behaviour—was not confined to humans, but then goes on to suggest that a new kind of "replicator" has recently emerged on the planet. This is the "meme", which is a unit of cultural transmission. The concept of the "meme", and the theory of memetics, is therefore closely associated with the work of Richard Dawkins.

Although Dawkins always distanced himself from Edward Wilson's sociobiology, the two men in fact shared a "gene's eye view" of biology, and Wilson had also suggested a similar concept to that of the meme: the culturegen (Lumsden and Wilson 1983: 121).

Memes, according to Dawkins, are—exactly like genes—"replicators"; they are cultural units of imitation, and spread from brain to brain. They propagate themselves, form "meme-complexes", and are subject to continuous mutation and blending. As examples of memes, he mentions "tunes, ideas, catch-phrases, clothing fashions, ways of making pots, or building arches". Additionally, Dawkins suggests that memes have the characteristics of longevity (they frequently exist in the brain for long periods), fecundity (they are easily copied and spread rapidly), and copying-fidelity (they can be exactly replicated) (1976: 191-195).

In later years, verily equating Darwin's evolutionary theory with atheism, Dawkins became increasingly hostile towards religion. He thus tended to view religious ideas, particularly the idea of God, as a "virus" inhabiting people's brains, a view consonant with his memetic theory (2006: 186). Perhaps Neo-Darwinian theory, as a meme-plex, is also a "virus"—or a "weed", to employ Blackmore's (1999: 41) imagery—inhabiting the brain of their "host"—Richard Dawkins!

Some twenty years after the publication of *The Selfish* Gene, the theory of memetics was given a new lease of life, in spite of the initial hostile criticisms of many social anthropologists. Two scholars are worth mentioning here, each the author of best-selling books.

One is the Neo-Darwinian philosopher Daniel Dennett, who strongly advocates a "meme's eye" perspective and graphically describes memetics in terms of the *Invasion of the Body Snatchers*. He thus views memes as "invading" or "infesting" the human person, and as being in "competition" with each other for entry into as many human minds as possible (1995: 34).

As with Dawkins, Dennett misleadingly equates Darwin's evolutionary biology with his own Neo-Darwinian gene-centred approach, suggesting that Darwin's dangerous idea (of natural selection) "is reductionism incarnate, promising to unite and explain just about everything in one magnificent vision" (1995: 82). Whether Darwin had such a one-track mind is certainly debatable (Midgley 2000: 73; Elsdon-Baker 2009: 126-127).

The other scholar is the Buddhist psychologist Susan Blackmore, who embraced memetics—Dawkin's theory of memes—with an unbounded enthusiasm. She viewed it as completely transformational in our understanding of the human mind. Around a hundred years after Wilhelm Dilthey and Franz Boas, she suddenly discovered the importance of human symbolic culture—atomized as memes—in the understanding of human life and human culture. With some pretension, Blackmore suggests that memetics, as a theory, will completely transform our mode of thinking in ways akin to the intellectual transformations associated with Copernicus and Darwin! (1999: 8).

The theory of memetics—the emphasis on the autonomy of cultural traits that seemingly "infest" their human "hosts"—is viewed by Blackmore as providing an explanation for a wide variety of social phenomena. "Memes" (i.e., culture), according to this scholar, have produced the large brains that humans possess, our capacity for language (the function of language is simply to spread memes), the nature of the self (as the "ultimate memeplex"), as well as explaining cultural evolution, which proceeds "in the interests of a selfish replicator"—a meme (1999: 30).

Anthropologists, of course, have stressed the importance of symbolic culture in the evolution of humans for many generations (Pfeiffer 1982; Dunbar et al 1999; Barnard 2012). Drawing on an earlier study (2014A: 97-99) I shall briefly enumerate some of the many criticisms that anthropologists have made with respect to memetics—the theory of the meme.

First, it is hardly an original theory. It is largely a revamping of diffusionist theory, long ago abandoned by anthropologists. What is new is that this theory has simply been yoked to Neo-Darwinian theory, with its emphasis on competition, fitness, and natural selection.

Second, it implies a form of cultural determinism and the kind of blank slate theory of the mind that Pinker had, rather ironically, so stridently critiqued. Recourse to the notion that cultural ideas—memes—are a kind of virus that infects human minds denies that human beings have autonomy, and through their associations can select and sustain their own beliefs, values, and ways of life. This denial of the autonomy and social agency of human beings has been described by some scholars as a rather "sinister aspect" of the meme's eye view of human life (Laland and Brown 2002: 229).

Third, many scholars have been critical of the "atomistic" perspective adopted by the meme theorists, emphasizing the problem of identifying a specific and distinctive unit of culture. What memeticists have suggested as an example for memes is, as Maurice Bloch suggests, a rather "ragbag" of proposals (2005: 92). But memes cannot be understood in isolation. For example, the idea of the "Virgin Mary" as a meme only makes sense in the context of the beliefs and practices that constitute Christianity, specifically the Roman Catholic tradition. This ideology (memeplex!), in turn, can only be understood if situated in a socio-historical context and in terms of specific social institutions, namely the church and various religious organizations (Morris 2006). To interpret cultural ideas as free-floating, "selfish" entities is, therefore, seriously misguided.

Fourth, many scholars have suggested that the emphasis on "imitation" and the notion that memes are simply "replicated" are both also misplaced, as they hardly accord with the complexity of cultural transmission or social learning. Cultural transmission is not simply a matter of a "meme" travelling from one brain/mind to another. It is more a process of the re-creation and reproduction, rather than the "replication", of information or cultural ideas. As recent scholars have concluded: "The learned, developmental dimension to the generation and reproduction of most cultural information makes it very difficult to think of cultural evolution in terms of distinct replicators or vehicles." (Jablonka and Lamb 2014: 207).

Finally, Dawkin's concept of "extended phenotype" (1982), adopted by both Dennett and Blackmore, hardly makes much sense when applied to humans. In fact, given the complex relationship that all organisms have with their environment, it has been suggested that the theory itself "explodes into caricatures" (Rose et al 1984: 273). But, in relation to humans, are we to conclude that every idea, belief, and ideology, every artefact—from pots and tractors to city landscapes and cultural environments—and every social institution and practice—from marriage rituals to capitalist corporations—are all part of the "extended phenotype" of the human species? It is hardly informative. (For useful criticisms of memetics see Sperber 1996: 100-106; Ingold 2000; Laland and Brown 2002: 224-232; Distin 2004; Bloch 2005: 87-101.)

Gene-Culture Co-Evolution

In an early critique of sociobiology, William Durham argued that Wilson, though claiming to offer a new synthesis between biology and the social sciences, in fact presented a rather one-sided account. He pointed out that Wilson focussed entirely on genetic inheritance mechanisms, thus ignoring the importance of the cultural mechanism as a mode of human adaptation. Durham thus postulated that human beings have two principal inheritance mechanisms, and that the cultural mechanism—cultural patterns and behavioural attributes that serve to enhance human adaptation and survival and are acquired through social learning—is not less important than the biological one. The human capacity for culture allows humans to modify aspects of the phenotype without any concomitant genetic changes, he suggested, and an adequate co-evolutionary theory must embrace both mechanisms. The process of "cultural selection" functionally complements that of natural selection (Durham 1979). It has been suggested that Durham did more than most scholars to pioneer "gene-culture co-evolution" as an empirical science (Laland and Brown 2002: 381).

Wilson appears to have acknowledged such criticisms. His study *Promethean Fire* (1983), co-authored with Charles Lumsden, advocated a theory of gene-culture co-evolution, although unlike Durham's early essays it emphasized the "linkages" between genes and culture. This implied, as they write, an "interaction in which culture is generated and shaped by biological imperatives while biological traits are simultaneously altered by genetic evolution in response to cultural innovation" (1983: 9).

But as I noted earlier, Wilson tended to see a "tight linkage" between genetic evolution and cultural history, mediated by "epigenetic rules of mental development"—specific universal psychological constraints. Throughout their study, Lumsden and Wilson focus on the relation between genes and culture, linked by a "great chain of causation" (1983: 70).

Many other scholars embraced the theory of "gene-culture co-evolution", including the pioneer geneticists Luca Cavalli-Sforza and Marc Feldman (1981), the anthropologists Robert Boyd and Peter Richardson (1985), and the aforementioned William Durham (1991). Recognizing that genetic evolution and cultural evolution—social change—are distinct processes, the "dual inheritance" theorists nevertheless indicated that there were important links between genes and culture. The most famous of these links, and one well-documented in the literature, related to the fact that people in both Europe and sub-Saharan Africa who possess a pastoral economy have developed lactose tolerance as adults, enabling them to consume cow milk and other dairy products without any ill-effects. This is seen as a "splendid" example of how culture can produce genetic change (Laland and Brown 2002: 260-262; Coyne 2009: 237-238).

There are two main criticisms of gene-culture co-evolutionary theory. The first is that it largely follows memetic theory in conceiving culture as consisting of separate, discrete units—as memes or culturegens. It is thus the equivalent of the "bean-bag" theory of genetics, and is open to the same criticisms that scholars have made of memetics—discussed above (Kuper 1994: 150; Bloch 2005: 87-101).

Second, in putting a focus on the extremes of gene and culture, and viewing the individual as simply a "carrier" of genes and memes, the theory of gene-culture co-evolution largely bypasses what is crucial to the understanding of human social life; namely, human development, agency, and the socio-historical context—the varied matrix of social relations in which people are engaged, including their relationship to the natural world. Indeed, culture is not so much an attribute of the individual, but rather a process intrinsic to a specific social group or collectivity (Morris 2014A: 102-103).

To conclude, sociobiology, and its various off-shoots—specifically evolutionary psychology, the theory of memetics, and gene-culture co-evolution—were all expressions of the Neo-Darwinian paradigm. The key figures in the development and popularity of this paradigm, with its gene-centred perspective, were undoubtedly, as I suggested, the biologists Edward Wilson and Richard Dawkins, well supported by the philosopher Daniel Dennett. This paradigm is often portrayed, especially by Dawkins, as if it were synonymous with Darwin's own theory of biological evolution. But this is quite misleading, as Darwin had a much more pluralistic approach to biological evolution (Elsdon-Baker 2009: 103-123).

I have discussed above some of the main criticisms of the Neo-Darwinian paradigm, with specific reference to sociobiology and its various offshoots. The theory has, of course, long been critiqued by various scholars from diverse intellectual traditions. Dawkins' arch-critic Stephen Jay Gould, in particular, emphasized that natural selection and evolution were not simply about genes; they were also about the individual organism—the basic unit of natural selection—and about groups of organisms—with respect to humans, their social life and their culture—as well as with the species. For Gould, species were real entities in the world and their very existence was dependent on wider geological forces (Gould 1983: 72-78, 2006: 438-460).

But although in this chapter I have tended to focus on—indeed, to stress—the critical limitations of sociobiology, we also need to recognize the more positive aspects of Neo-Darwinian theory, namely its embrace of philosophical materialism and its strong advocacy of an evolutionary approach to biology as well as to human social life and culture. It is thus a healthy antidote to the subjective idealism and textualism of much of the research generated by postmodern anthropology. Unfortunately, given its obsessive gene-centred focus, sociobiology, as many scholars have stressed, is essentially a form of reductive and rather mechanistic materialism (Bookchin 1995B: 39; Bunge 2001: 74).

One should not, of course—as some scholars do—confuse reductionism (or physicalism) with analysis, which is an integral part of the scientific method or any dialectical (or relational) epistemology.

[It is worth noting that in the past decade critiques of Neo-Darwinism have almost become a culture industry; see, for example Tallis 2011; Tudge 2013; Fernandez-Armesto 2015.]

Chapter Seven

Marx's Dialectical Materialism

"The history of the twentieth century is Marx's legacy", wrote the acclaimed biographer of Karl Marx (Wheen 1999). Given that Josef Stalin and Mao Zedong both claimed to be his heirs, this may well be true. Yet it hardly adds to our understanding of one of the great intellectual figures of the nineteenth century. In any case, as Wheen acknowledged, Marx would undoubtedly have repudiated the politics and tyranny associated with the Soviet Union under Stalin and the Chinese State under Mao. Both were forms of State Capitalism under an oppressive party dictatorship, far removed from Marx's embrace of democratic politics and his conception of a Communist society. As Sidney Hook succinctly put it: "Marx was a democratic socialist, a secular humanist, and a fighter for human freedom. His words and actions breathe a commitment to a way of life and a critical independence completely at odds with the absolute rule of the one-party dictatorship of the Soviet Union" (1971: 2)—or any form of dictatorship.

Only a few years ago, Marxism was seen as being at a very low ebb, presented as having "an inglorious past and no future" (Sheehan 1985: xv). Apologists for global capitalism like Vernon Bogdanor (2003) described Marx as a "relic" from the past, and Marxism as a religious cult that had no contemporary relevance. However, since the demonstrations against the World Trade Organization in Seattle in 1999, there has been a resurgence of interest in Marx, both as a social theorist and as a major critic of the capitalist economy. The literature on both Marx and Marxism is therefore now vast. I have no intention of reviewing this literature here, but will focus specifically on Marx's own philosophy, his "way of thinking about the world", which is well described as *dialectical materialism*. This term, rather unfortunately, has tended in the past to be identified with the philosophical musings of Josef Stalin. However, in recent years it has rightly been reclaimed by Marxist scholars as a true, and appropriate, description of Marx's (and Engels') own philosophy (Woods and Grant 1995: 43-79; Foster 2000; Foster et al 2010: 249-287).

In this chapter I will focus specifically on three topics: Marxist dialectics as a way of understanding the world, Marx's own conception of a materialist philosophy, and, finally, dialectical biology, a current of thought which arose at the end of the last century as a critique of the Neo-Darwinian paradigm of the sociobiologists.

A Dialectical Mode of Thinking

In his early writings, Karl Marx was essentially engaged in combining Hegel's philosophy, with its emphasis on the historicity of being and its dialectical way of understanding the world—while rejecting its spiritualist idealism—with the philosophical materialism of Ludwig Feuerbach. For Feuerbach's ontology placed a crucial emphasis on our earthly experience, and on the human subject as a living being rather than—as with Hegel—on the vicissitudes of *geist* (as spirit or universal mind).

Marx always and continually paid tribute to Hegel as a dialectical thinker. It is therefore, I think, somewhat misleading to view Hegel as a "monkey hanging around Marx's neck" (Harris 1980: 145), or to say that Marx made a radical epistemological break or leap from ideology (idealism) to science by renouncing Hegel's historicism and his dialectics (Althusser 1977: 13-14; Morris 2014A: 645-646). Marx never did renounce dialectics, but rather incorporated it into the materialist philosophy and his own understanding of science.

Engels in his later years was to stress the importance of Hegel's philosophy and the dialectical outlook that both he and Marx embraced. He thus described Hegel's philosophy as "epoch-making" and, acknowledging that Hegel had a truly encyclopedic mind, Engels wrote: "For the first time the whole world, natural, historical, intellectual, is represented as a process ie [sic] as in constant motion, change, transformation, development." (Engels 1969: 34). What Marx and Engels repudiated was not Hegel's historicism nor his dialectics, but rather his metaphysics—his "pantheistic mysticism" (Marx 1975: 61).

A good deal has been written on the concept of dialectics. Some have dismissed the notion as a form of mystical mumbo-jumbo. Sydney Hook asked what Galileo's laws of motion and the life-history of an insect have to do with dialectics, and responded, like many critics of Engels, by suggesting that dialectics was only applicable to human social life (1971: 75-76). Likewise, Mario Bunge considered dialectics to be an essentially obscurantist, unhelpful legacy of Hegel, although he acknowledged the importance of Marx and Engels as materialist philosophers and pioneer social scientists (1999B: 133).

What then is the "dialectical method" as conceived by Marx, and Engels? To answer this question, it is perhaps best to turn to the writings of Frederick Engels, whose own philosophical and intellectual interests were extremely wide-ranging. Engels, in fact, was especially fascinated by the developments within the natural sciences towards the end of the nineteenth century. He was particularly excited by three important developments, namely, the emerging theories of thermodynamics, concerning the conservation of energy (Hermann Helmholtz); the discovery of the cell-structure of all living organisms (Rudolf Virchow); and the new metaphysics of nature that had been heralded by Darwin's evolutionary theory (Marx and Engels 1968: 610).

Engels thus conceived of dialectical thought as entailing a materialist conception of nature and of human history that was directly based on these new scientific discoveries. He saw this as implying a new conception of nature and a new mode of thought—dialectics—that was directly opposed to what Engels described as "metaphysics"—a term covering both Hegel's idealism and the static Newtonian conception of the universe that he referred to as "mechanical materialism". For Engels, then, "dialectics" essentially implied three principles: an emphasis on process, temporality, and change; a conception of totality that implied a focus on relations and correlations; and, finally, a stress on "contradiction". I shall briefly outline these three principles in turn.

Engels has often been portrayed as a crude positivist or as a mechanistic materialist. This is not only extremely unfair to Engels but displays a woeful misunderstanding of his work. For Engels was acutely aware that the scientific revolutions of the nineteenth century, especially that of Darwin, had completely transformed our understanding of nature and human life. These developments suggested that "Nature also has its history in time" (1969: 35).

Thus the first principle of dialectics is the view, expressed long ago by the Greek philosopher Heraclitus, that all things in the universe are in a process of change. Nature, consequently, is historical at every level, and no phenomenon of nature simply exists: it has a history, it comes into being, it endures, it changes or develops, and finally it ceases to exist. Aspects of nature may appear to be fixed or stable, or in static equilibrium, but nothing is permanently so. As Engels expressed it, citing Heraclitus, "everything is *fluid*, is constantly changing, constantly coming into being and passing away" (1969:30). Or, as he expressed it elsewhere: "the great basic thought [is] that the world is not to be comprehended as a complex of ready-made things, but as a complex of processes in which the things, apparently stable, no less than their mind-images in our heads, the concepts, go through an uninterrupted change of coming into being and passing away" (Marx and Engels 1968: 609).

The first principle of Engels' understanding of dialectics is then the idea that both the natural world and human social life are in a constant state of flux, and that modern science has made the "immutable" concepts of nature held by Newton, Linnaeus, and Hegel completely redundant. Thus long before Bergson and Whitehead—and such contemporary luminaries as Deleuze, Capra, and Judith Butler—Engels emphasised the importance of "becoming", stressing in embryonic form of a procedural—dialectical—philosophy, one that emphasized that the world was not a spiritual entity (Hegel), nor a machine (Newton), but a historical material process.

The second principle in Engels' understanding of dialectics emphasized the notion of totality. This is the idea that all the seemingly disparate elements of which the world is constituted are interconnected and that no phenomenon (whether physical, biological, or social) can be fully understood in isolation, but rather must be seen as a part of a complex totality. This principle entailed an evolutionary (or dialectical) form of holism and a conception of nature that was neither cosmological nor mechanistic but instead ecological (Morris 1981). Dialectics, Engels wrote, "comprehends things and their representation in their essential connection", and he emphasized the importance of Darwin's theory of evolution that had dealt a critical blow to the "metaphysical" conception of nature (i.e., mechanistic philosophy) in showing that all organic beings—plants, animals, humans—are the products of evolution and thus interconnected (1969: 33).

The emphasis on totality is by its very nature opposed to any form of reductionism—of explaining the whole by means of its parts. Nor, it must be stressed, does it abolish the autonomy and role of its parts (the individual component) in favour of the whole—the totality. Engels was not advocating a teleological or mystical form of holism, as expressed, for example, in the writings of Jan Smuts (1926; for an important critique of Smuts' idealistic holism see Foster et al 2010: 315-324).

Thus, a hundred years before deep ecology and the advocacy of a "systems view of life" (Capra and Luisi 2014), Engels emphasized the crucial importance of connectedness, relationships, and context, and the fact that all things are interrelated and interdependent as well as forming complex wholes.

The third principle of dialectics was expressed by Engels in terms of the notion of "contradiction" or the "unity of opposites". Ordinary common-sense understanding, formal logic, and metaphysical philosophy (especially as expressed by Descartes and Kant), tended to imply, Engels suggested, thinking in terms of "absolutely irreconcilable antitheses". Hegel had referred to this mode of thinking as understanding (*verstand*), thinking in terms of dualistic oppositions. Engels described it as "metaphysical": as an abstract, restricted mode of thought, one lost in "insoluble contradictions". He maintained that certain oppositions—cause and effect, appearance and essence (reality), identity and difference, freedom and necessity—"mutually interpenetrate" and are best conceived as a "unity in opposition". This is particularly the case in the alleged opposition between humanity and nature, and between the human individual and the societies to which they belong. Engels further argued that a characteristic typical of processes of change is the "negation of the negation"—the development of a new synthesis that negates, preserves, and transcends (*aufheben*) the elements of the contradiction. Engels therefore made a clear distinction between two philosophical tendencies: "the metaphysical with fixed categories; the dialectical (especially Aristotle and Hegel) with fluid categories" (1940: 153).

Engels summed up his critique of metaphysics—specifically mechanistic philosophy—by suggesting: "In the contemplation of individual things, it forgets the connections between them: in the contemplation of their existence, it forgets the beginning and end of their existence; of their repose, it forgets their motion." (1969: 32). Engels thus concluded that "modern materialism is essentially dialectic" (1969: 36). In the simplest terms, he wrote that dialectics was "nothing more than the science of the general laws of motion and development of nature, human society and thought" (1969: 169).

Engels'basic contention then was to emphasize the importance of historicism (historical understanding in the widest sense), that science itself was a historical mode of thinking, and that the familiar dichotomies—identity/difference, chance/necessity, body/mind, subject/object, humanity/nature, and individual/society—must not be viewed as irreconcilable opposites or antithesis but as dialectically interrelated, as expressing a "unity in opposition".

The efforts that Engels made to apply the "dialectical method" to both nature and human society has been described as "heroic"— it was dialectically consistent with both a materialist outlook and with the important developments in nineteenth-century science. Thus Engel's dialectics of nature can be viewed as a doctrine of emergent evolution, as a "genetic-historical approach rooted in a philosophy of emergence" (Foster et al 2010: 2; for further studies of Engels' dialectics of nature see McGarr 1994; Sayers 1996; Callinicos 2006: 209-216).

The Marxist geographer David Harvey, though also a strong advocate of "dialectical thinking", is more critical of Engels' approach and emphasized that dialectics cannot be "imposed" on the world. But Harvey quite misleadingly suggests that dialectics give "priority" to processes (and relations) over things, thus denying their dialectical relation and the agency of organisms (Harvey 2016: 196-199).

Let me now turn to Marx's (and Engels') materialism.

Marx's Historical Materialism

Marx is famous, of course, as a social scientist, specifically for his analysis of the capitalist economic system, embodied in his classic work *Capital* (1867). In a seminal preface to his critique of political economy, in what he described as the "guiding thread" for his own studies, Marx outlined what came to be known as the materialist conception of history. Marx wrote:

> *In the social production of their life, men enter into definite relations that are indispensable and independent of their will, relations of production that correspond to a definite stage of development of their material productive forces. The sum total of these relations of production constitutes the economic structure of society, the real foundation, on which rises a legal and political superstructure, and to which correspond definite forms of social consciousness. The mode of production of material life conditions the social, political and intellectual life process in general.*

Marx continues with that famous phrase: "It is not the consciousness of men that determines their being, but, on the contrary, their social being that determines their consciousness" (Marx and Engels 1968: 181).

Marx's materialist conception of human social life (history) has given rise to a welter of literature, much of it critical, implying that Marx was a crude economic determinist. But although Marx speaks of material life as "conditioning" or "determining" other aspects of socio-cultural life, it is, I think, misleading to interpret this as implying a simple, direct causal relationship between the economic base and the political and cultural superstructures. To do so is to invoke a mechanistic paradigm which is quite alien to Marx's tenor of thought. As Merleau-Ponty put it: the economic base is not a "cause" but the "historical anchorage" for law, religion, and other cultural phenomena (1964: 108-112).

But what must be recognized is that Marx's materialist conception of (human) history, as a social theory, has to be situated within his wider philosophical worldview—dialectical (historical) materialism. Indeed, Leszek Kolakowski (1978) began his important history of Marxism with the words "Karl Marx was a German philosopher". Although Marx was also an economist and a revolutionary socialist scholar, with an almost encyclopaedic knowledge of many fields, he was fundamentally a philosophical materialist. Marx's materialist outlook was essentially derived, or at least confirmed, with respect to three main sources.

The first were the ancient Greek philosophers Democritus and Epicurus, whose writings form the basis of Marx's own doctoral thesis (1841). His notes for this thesis he titled *Epicurean Philosophy*. Both Greek scholars were ardent materialists (and atheists). Yet while the atomist Democritus placed a strong emphasis on necessity (determinism), the more empirical Epicurus stressed the importance of contingency and human freedom (for Marx, of course, chance and necessity were both aspects of the material world, and dialectically interrelated). It has therefore been suggested that what Marx derived from Epicurus was a sophisticated form of materialism, one that rejected the concept of God and the teleological principle of religion; that stressed the importance of sensual experience while also acknowledging the role of the human intellect in the assessment of empirical data; and, finally, that emphasized the importance of contingency. For Marx, Epicurus heralded, in embryonic form, a non-mechanistic, non-deterministic form of materialism. But what particularly appealed to Marx was that Epicurus initiated a philosophy of human freedom. In fact, according to Epicurus, to serve philosophy was to seek human freedom. It was this liberating aspect of Epicurus that Marx most admired, and he (and Engels) often paid tribute to Epicurus as the "greatest figure of the Greek Enlightenment" (Marx and Engels 1965: 150; McLellan 1973: 34-38; Foster 2000: 51-65, 2009: 150).

The second important influence on Marx was the Bavarian scholar Ludwig Feuerbach. There is no doubt that Feuerbach's philosophy—his critique of religion and of Hegel's metaphysical idealism—had a profound impact on Marx and Engels. As Engels described it, Feuerbach's study *The Essence of Christianity* (1841) "placed materialism on the throne again". It was greeted with enthusiasm by Marx and Engels, and had a real "liberating effect" on their way of thinking (1968: 592). But they were, nevertheless, highly critical of Feuerbach's rather abstract conception of the human subject and his rather ahistoric materialism.

In his famous *Theses on Feuerbach* (1845), Marx described Feuerbach's philosophy as a form of contemplative materialism which conceived the relationship of humans to the natural world (reality) only in terms of thought or theory, not in terms of humans' sensuous, practical activities with regard to nature (Marx and Engels 1968: 28-29). Feuerbach thus failed to acknowledge the significance of the active and interactive relationship that humans have with the natural world. For Marx, in contrast, there was an intrinsic organic link between humans and the natural world, such that life itself emerges from the interactions between an organism and its natural environment. All life, as one Marxist scholar has written, "is based on metabolic processes between organisms and their environment" (Foster 2009: 50). Indeed, Marx described the labour process itself—our means of obtaining our basic livelihood—as essentially one involving a complex metabolic interaction between the human organism and the material world.

As Marx wrote: "Labour is, first of all, a process between man and nature, a process by which man, through his own actions, mediates, regulates and controls the metabolism between himself and nature. He confronts nature as one of her own forces.... the (labour process) is the universal condition for the metabolic interaction between man and nature, the everlasting nature-imposed condition of human existence" (Marx 1957: 169-177; Foster 2000: 157).

It was this practical engagement or metabolic interaction between humans and the earth that Marx felt had been completely neglected by Feuerbach, as well as the latter's failure to understand the historicity of nature. Thus Marx and Engels concluded: "As far as Feuerbach is a materialist he does not deal with history, and as far as he considers history he is not a materialist. With him materialism and history diverge completely" (1965: 59-60).

Aiming to combine Feuerbach's materialism and his emphasis on the human subject with Hegel's historicism and relational epistemology, Marx and Engels were insistent on the essential unity of nature and history. Social life and culture were, for Marx and Engels, intimately linked with the natural world. On these same grounds they were therefore equally critical of the left-Hegelian Bruno Bauer, for Bauer tended to treat history and nature as antithetical concepts. Marx and Engels stressed that humans have both a "historical nature" and a "natural history", that the celebrated "unity of man with nature" had existed throughout history in relation to human productive activities, and that "nature" was being continually transformed by human activity. Marx and Engels thus wrote: "The nature that preceded human history was by no means the nature that Feuerbach encountered" (1965: 58-59; for useful discussions of Feuerbach's philosophy see Kamenka 1970; Althusser 2003: 85-154).

The third important influence on Marx's materialist outlook was, of course, Charles Darwin. As I indicated earlier, Marx and Darwin, in philosophical terms, were kindred spirits. It is well-known that Marx, on his initial reading of the *Origin of Species* (1859), declared it to be "absolutely splendid", and considered it an "epoch-making work". For it fully demonstrated, he felt, that there was "historical evolution in nature." Darwin's evolutionary theory, Marx contended, not only presented the "death blow" to any "teleology" in nature, but also formed the "natural history foundation" for Marx's own theory of human history (Padover 1979: 139; Patterson 2009: 87).

Yet Marx also reflected on the fact that, in applying Malthusian theory to animals and plants, Darwin—as Marx wrote in a letter to his friend Engels (June 1862)—discovered "among beasts and plants his (own) English society with the division of labour, competition, opening of new markets, and the Malthusian "struggle for existence". It is Hobbes' *bellum onmia contra omnes* (war of all against all) (Padover 1979: 157). Marx was thus intimating the degree to which Darwin's theory of organic life mirrored capitalist social relations that were developing in nineteenth-century Britain.

It is, however, important to recognize that Darwin and Marx had much in common. Both unreservedly embraced philosophical materialism, rejecting all forms of transcendence, whether expressed as idealist philosophy or as a religious or spiritualist metaphysic. Both developed a ratio-empiricist form of science, a relational epistemology (dialectics) that stressed the importance of both empirical knowledge and rational thought. And, finally, both were historicists, stressing the need for historical understanding and thus envisaging a historical (dialectical) science. For, as Lynn Margulis put it, "Evolution is history." (Margulis and Sagan 1971: xxiii). Darwin and Marx differed in that Darwin's evolutionary naturalism focused largely, but not exclusively, on organic life, while Marx's historical materialism was essentially a social theory, concerned with the understanding of human social life, specifically the capitalist economy.

It is therefore important to recognize that both Marx and Engels were naturalists—in a philosophical sense—and realists. They were realists in firmly acknowledging the objective reality of the material world. As Engels wrote: "Nature exists independently of all philosophy." (Marx and Engels 1968: 592). They were also naturalists in the suggestion that there is only one world: the world of nature. There is no absolute spirit or deity; neither is there any separate spiritual world or spirit entities of any kind. As Engels argued in his famous essay on "Feuerbach and the End of Classical German Philosophy" (1888), the material, perceptible world to which humans belong is "the sole reality" and the mind (spirit/consciousness) is merely "the highest product of matter". As Engels put it: "Nothing exists outside nature and man" (Marx and Engels 1968: 592-596; Foster 2000: 77).

Two points may be made here. The first is that Marx and Engels' conception of materialism is very different from the use of the concept in contemporary academic philosophy, where materialism invariably indicates a form of reductive materialism, or what Marx, and Engels would have described as "mechanical materialism" (c.f., Moser and Trout 1995). The second is that, although critiquing the subject/object dualism of Cartesian philosophy—what Engels described as metaphysical thinking—neither Marx nor Engels responded by completely dissolving this distinction into some woolly abstraction like "experience" or "meshwork". Even less did they envisage collapsing the distinction between humans and nature into the intuition of some mystical unity—such as the "one" of Plotinus or Advaita Vedanta. They rather emphasized that humans (as concrete organisms) and the natural world were interdependent and dialectically interrelated (Morris 2014A: 20-21).

Many years later the Brazilian scholar Paulo Freire was to express the essence of Marx's materialism, saying: "To be human is to engage in relationships with others and with the world. It is to experience that world as an objective reality, independent of oneself, capable of being known" (1974: 3).

Within the Marxist tradition there has long been an ongoing debate, and at times harsh polemical exchanges, between two distinctive interpretations of Marx.

On the one hand, there are those usually described as *critical* or Hegelian Marxists, who stress the continuity of Marx with Hegel and who view Marxism as critique rather than as a science. They thus take a more "historicist" or "humanistic" interpretation of Marx's writings and situate themselves in the more literary and philosophical tradition of European culture. Often they are highly critical of science and technology. Scholars such as Lukacs, Gramsci, Sartre, Adorno, Marcuse, and Fromm are usually placed within the critical Marxist tendency.

On the other hand, there are the *scientific* Marxists, who stress that Marxism is a science of history and suggest that Marx made a clean break with Hegel's historicist philosophy. Among the early writers, Kautsky and Plekhanov are placed within this tradition, along with Engels, and they are held to present a more deterministic and positivistic interpretation of Marx's writings. Among contemporary scholars who eschew critical theory and historicism, mention may be made of Godelier, Poulantzas, and Althusser, as well as the cultural materialist Marvin Harris. With Louis Althusser a radical break was stressed between the young Marx, still allegedly enrapt in Hegelian ideology, and the mature Marx, the text of capital being seen as an exemplification of true science. Althusser thus expressed a strident anti-humanism.

At extremes, the first tendency degenerates into naïve romanticism and hermeneutics, whereas the second slides into positivism and mechanistic science. But, as we have indicated above, what was important about Marx—the essence of his contribution to materialist philosophy—was that he consistently tried to unite these two opposing tendencies, advocating an approach that was both materialist (scientific) and dialectical (historical): dialectical materialism.

This was, of course, long ago emphasized by scholars such as Sebastian Timpanaro (1975) and George Novack (1978). Both these scholars offered a sterling defence of Engels' contribution to Marxist philosophy, against the crude depiction of Engels as a vulgar materialist offered by Lukacs (1971) and by a host of other Marxist scholars (e.g., Lichtheim 1971; Schmidt 1971; but c.f. Sheehan 1985: 53-60 for an illuminating discussion). What particularly troubled Timpanaro was that the majority of Western Marxists (invariably university professors), particularly Lukacs and the critical theorists (Adorno, Horkheimer), strongly influenced as they were by neo-romanticism and phenomenology, had essentially produced an "anti-materialist" version of Marxism. Ignoring the relationship between humans and nature (which was crucial to Marx), many Western Marxists seemed to deny the conditioning which nature continually exerts upon human life, and virtually denied the biological aspects of human life (Timpanaro 1975: 10).

On the other hand, Althusser and the structural Marxists, along with the anthropologist Levi-Strauss, had gone to the other extreme. With "stylistic prosperity" and "theoretical pretension", they had advocated a form of Marxism that eliminated the human subject entirely from social analysis, was deeply anti-historical, and, given its focus on "pure theory", largely rejected the empirical world and lived experience. Timpanaro indeed described structuralism as a rather mystifying platonic scientism (1975: 186-187).

More recent scholarship has essentially affirmed what Timpanaro and Novack suggested many years ago, namely that Marx's philosophy was a unique synthesis that combined a comprehensive materialist worldview with a theory of universal evolution that was thoroughly dialectical. Novack emphasized that Marx's view of both nature and human social life was profoundly historicist, and, like Spinoza and Hegel, Marx expressed a unitary conception of being. To suggest that Marx was a Cartesian dualist with hatred and contempt for the material world (Popper 1945: 102-103), or to interpret Marxism in a Neo-Kantian fashion (Lukacs 1971), is therefore quite misleading. For Marx, nature and human social life form two contrasting parts of a single historical process, and just as there is both continuity and discontinuity in the evolutionary transition from ape to human, so there is a compatible continuity and discontinuity between the dialectics of nature and those of human history (Novack 1978: 244; Morris 2014A: 31-37).

It may therefore be concluded that the form of materialism that both Marx and Engels advocated was one that was realist, critical, historical, and dialectical, as well as expressing an ecological mode of thought (Foster 2000, 2009: 143-160). In a real sense, both these radical scholars were true heirs of the Enlightenment, advocating an uncompromising materialism "which embraced such concepts as emergence and contingency and which was dialectical to the core" (Foster 2009: 153).

Dialectical Biology

The anticipated demise of Marxist philosophy—heralded or at least hoped for by both neoliberals and neoconservatives alike, along with the alleged "end of history" (Fukuyama 1992; Ryan et al 1992)—never seems to have materialized. In fact, in response to the reductive materialism of Neo-Darwinian theory, discussed in the last chapter, Marxism as a worldview was given a new lease of life towards the end of the last century. This rejuvenation of Marxist scholarship was particularly associated with a coterie of scholars who developed what has been described as "dialectical biology". These scholars attempted to articulate a way of understanding the natural world and human social life that was very different from both the mechanistic materialism of the Neo-Darwinian paradigm and the kind of mystical holism that was particularly associated with the ecologist Frederic Clements (1949) and the popular writings of some spiritual ecologists (Capra 1982). The group included scholars from diverse intellectual backgrounds, namely the palaeontologist Stephen Jay Gould, the population biologist Richard Levins, the geneticist Richard Lewontin, and the neuroscientist Steven Rose. All were critical of what they described as "Darwinian fundamentalism", given its reductive materialism and its gene-centred perspective, as well as the "obscurantist holism" of mystical ecologists (Levins and Lewontin 1985: 133).

Genial, erudite, and iconoclastic, what made Stephen Jay Gould famous, at least in the United States, was not that he was adjudged to be a notorious communist—he was not in fact a Marxist but a radical liberal—but that he was a superb essayist. For around twenty-five years, beginning in 1974, Gould wrote over three hundred essays, offering his insightful reflections on natural history topics. The essays were republished in such well-known books as *Ever Since Darwin* (1980) and *The Panda's Thumb* (1983). An outstanding essayist in the genre of popular science, even his critic Richard Dawkins commended Gould's writings for combining the naturalists' love of life with a historians' respect and affection for her or his subjects (2003: 202).

In these popular writings, Gould presented many important criticisms of the Neo-Darwinian paradigm, as represented by Dawkins, Wilson, and the philosopher Daniel Dennett. Two criticisms are worth highlighting here.

The first is that Gould was critical of the Neo-Darwinists in that they virtually equated the evolution of life on Earth with the theory of natural selection. Gould, of course, fully endorsed Darwin's theory of evolution as descent through modification as well as his suggestion that natural selection is the most important mechanism of evolution. As Gould affirmed: "Natural selection, an immensely powerful idea with radical philosophical implications, is surely a major cause of evolution as validated in theory and demonstrated by countless experiments" (2006: 440).

But unlike the Neo-Darwinists, Gould did not consider natural selection to be the exclusive cause of evolutionary change. He frequently cited Darwin to support his own more pluralist approach to evolution, for in the introduction to the *Origin of Species* Darwin had concluded that natural selection had been the most important, but not the exclusive, means of modification. Among other mechanisms or factors that Gould, along with other biologists, have seen as "adjuncts" to natural selection are the following: horizontal gene transfer (especially among bacteria), the process of random genetic drift, and such factors as historical contingency and the importance of species in the evolutionary process (Gould 2006: 438-444).

Secondly, Gould was highly critical of the reductionism that was implied in the "genes' eye" view of evolution advocated by Wilson and Dawkins. Following in the footsteps of Ernst Mary (1988), Gould emphasized the crucial importance of two key concepts, namely, historical contingency (which does't imply a lack of cause) and emergence.

The concept of emergence can be effectively illustrated by the following oft-quoted example: although sodium is a poisonous metal and chlorine is a poisonous gas, in combination as sodium chloride—common salt—the latter has properties entirely different from its component parts (Pringle 2009: 62). Thus with regard to complex systems at whatever level of organization—genes, cells, organisms, deems (local populations), or species—the concept of emergence, Gould argued, was highly relevant. Organisms, for example, have properties that cannot be understood simply from a knowledge (however complete) of genes and cells (Gould 2003: 202). Gould therefore was highly critical of the reductionism implicit in the Neo-Darwinian paradigm, and continually emphasized the importance of both emergence and historical contingency. The gene, Gould felt, was not the only level at which natural selection occurs—as emphasized by Darwin—for selection takes place at all levels of biological organization: genes, organisms, deems (local groupings), and species.

What Gould was particularly concerned to avoid were two extremes: either setting up a "picket fence" between the human species and the rest of nature—a strategy clearly evident among post-modern or cultural anthropologists—or ethnocentrism, the reductive strategy of the sociobiologists, such as Wilson and Dawkins, who tended to reduce human social life to biology. Thus, unlike the sociobiologists and evolutionary psychologists, Gould recognized—like Mumford and Fromm—the inherent paradox in human life. As he wrote: "We live in an essential and unresolvable tension between our unity with nature and our dangerous uniqueness. Systems that attempt to place and make sense of us by focusing exclusively either on the uniqueness or the unity are doomed to failure" (Gould 1984: 250).

Equally significant, in the present context, is that Gould made a strong plea for a historical science. Noting that modern science had tended to denigrate history, Gould emphasized that many of the sciences of nature-cosmology, geology, palaeontology, and evolutionary biology were *historical* sciences. It was, he argued, misleading to view the scientific method as necessarily involving experiments, prediction, and reductionism, for the historical sciences combine determinism with a stress on contingency, and place an important emphasis on historical explanations (Gould 1989: 279). Of course, it must be noted, as I shall stress, that anthropology is also a historical science.

Yet it is interesting to note, given the fact that he always distanced himself from Marxism, that Gould in his writings rarely mentions the concept of dialectics. This, however, was a key concept for his friend and colleague Richard Lewontin, a distinguished geneticist and a committed Marxist. In 1984, together with Steven Rose and Leon Gamin, Lewontin produced the ground breaking text *Not in our Genes*, which critiqued not only sociobiology but all forms of biological determinism, whether in regard to race and gender, or with respect to the human subject more generally. Around the same time, Lewontin collaborated with Richard Levins, who had a strong interest in ecology and public health, to produce another important text: *The Dialectical Biologist* (1985). The text aimed to provide a Marxist or dialectical materialist worldview, in contrast to what they described as "Cartesian reductionism" and the kind of mystical holism that was then popular with deep ecologists and new-age enthusiasts.

The Cartesian philosophical worldview, though implying a radical materialist ontology, reflected the essential standpoint, Lewontin and Levins contended, of ultra-Darwinism, as expressed by Wilson and Dawkins. Indeed, they suggest that this worldview forms the preconceptions and the dominant mode of analysis not only of the physical and biological sciences, but also by extension of the social sciences. They critique this worldview—that of Neo-Darwinian theory—as implying an epistemology that was mechanistic, reductionist, and atomistic, interpreting it as the most recent form of positivist ideology (1985: 269-271).

But, as dialectical biologists, Levins and Lewontin were also critical of what they describe as idealist or obscurantist holism, which views the whole as the embodiment of some ideal or mystical organizing principle. This was expressed in Frederic Clements' (1949) concept of a plant community as a "super organism", and in Taoist holism, a form of holism that emphasizes balance and harmony rather than historical development (Levins and Lewontin 1985: 275).

Dialectical biology is therefore viewed by Levins and Lewontin as a form of dialectical materialism that negates both the mechanistic materialism and the genetic reductionism of the sociobiologists, on the one hand, and holism, the quasi-religious metaphysics which emphasizes only connection and balance, on the other (Lewontin and Levins 2007: 126).

Levins and Lewontin thus acknowledge, like Gould and many earlier scholars, the essential paradox of human life. They side with the evolutionary biologists in insisting on the intrinsic continuity between humans and other life-forms, while agreeing with the cultural idealists and holistic scholars in emphasizing the discontinuity between human social life and culture and the animal world, thus stressing human uniqueness (Levins and Lewontin 1985: 133).

It is significant that the study *The Dialectical Biologist* is dedicated to Frederick Engels. A century after Engels, Levins and Lewontin outline their own conception of dialectics as a mode of understanding. They suggest the essential tenets of dialectical thought can be expressed in the following brief statements:

- The truth is the whole, and all parts are conditioned, or even created by the whole. All things are more richly connected than is obvious, and all aspects of the physical world are in interaction with each other in some degree.

- No level of phenomena—gene, organism, species, ecosystems—is more "fundamental" than any other, and each has a relative autonomy and its own dynamic. At each level things are internally heterogeneous.

- Things are essentially snapshots of processes, and change is characteristic of all systems.

- The various dichotomies by which we understand the world—subject/object, organism/environment, contingency/determinism, mind/body, biology/social, nature/culture—are in a sense misleading, and potentially obfuscating, for they must be viewed as *dialectically* interrelated (Levins and Lewontin 1985: 272-275; Lewontin and Levins 2007: 149-150).

Lewontin and Levins recognized, of course, that we make distinctions between, and separate as mental constructs, the biological from the social, the body from the mind, subject and object, cause and effect. We do so all the time, and we have to in order to recognize and investigate phenomena. This analytical step, they write, is a "necessary moment in understanding the world". But to view them as strictly dichotomous—as opposed concepts—is misleading, reflecting what Engels described as metaphysical thinking. These oppositions must be viewed dialectically, and we need to conjoin them to show "their interpenetration, their mutual determination, their entwined evolution, and yet also their distinctiveness" (Lewontin and Levins 2007: 106).

It is important to recognize that the dialectical biologists made an important distinction between nature and the environment. There is indeed an external world that exists independently of any living being, and the totality of this natural world should not be confused with an organism's environment. For organisms, including humans, determine by their life-activities what is relevant to them. An example, they suggest, is that although a stone may be an essential part of the environment of a song thrush—as it is used as an anvil to break snail shells—the stone has very little relevance for a blue tit or woodpecker (Lewontin and Levins 2007: 231).

To illustrate the dialectical approach, and the "interpenetration" of the organism and its environment, Lewontin and Levins suggest the following: that organisms select their environments and modify them in several ways; that organisms do not passively adapt to the world, but creatively respond to the environmental factors that may impinge on their being; that the reciprocal interaction of the organism and its environment takes place through several pathways; and, finally, that organisms, including humans, both make and are made by the environment, and are thus actors in their evolutionary history (Levins and Lewontin 1985: 274). Lewontin and Levins thus conclude that "remaking the world is a universal property of living organisms"—including, of course, humans (2007: 234).

In the last two chapters, I have explored some of the many critiques of the Neo-Darwinian paradigm. The alternative perspective that was implied essentially suggested a return to Darwin and Marx's materialism. From Marx was derived the importance of a dialectical approach to the understanding of both biology and human social life, in contrast to the reductive materialism of sociobiology. From Darwin, we learn the need to shift the emphasis away from the "gene-centred" perspective of the Neo-Darwinian scholars, putting the focus back on the organism in its environment. But, of course, the organism (a category which includes humans) has to be conceived not as a "survival machine" or robot, through which genes (Dawkins) or "mimes" (Black more) replicate, but rather as functioning, autodidactic agents actively engaged in their own destiny (Rose 1997).

The gene-centred approach of the Neo-Darwinian sociobiologists and the tendency of some anthropologists (Inhold 2013A: 10), in reaction, to completely oblate the distinction between the organism (including the human person) and their natural environment (along with the subject/object distinction) are *both* untenable. The truth—the dialectic—was well expressed by the philosopher D.M. Walsh when he concluded that: "Evolution is an ecological phenomenon. It arises out of the purposive engagement of organisms with their conditions of existence" (2015: 246).

This cardinal idea of Charles Darwin seems to have been forgotten by both the gene-centred Neo-Darwinian biologists and both the postmodern and neo-structuralist anthropologists, ever eager to "dissolve" the subject—as Levi-Strauss expressed it (1966: 247)—with abstract relation (or lines!).

Both Marx and Darwin have had, of course, a profound impact on Western culture. With respect to the present manifesto, what has to be stressed is that both scholars were fundamentally historical thinkers, as well as being unabashed philosophical materialists. They, therefore, in different ways, embraced both historicism (in its broader sense) and naturalism.

I shall turn now to the third form of contemporary materialism, which places an important emphasis on the concept of emergence and which has been variously described as the "systems view of life", or complexity theory.

Chapter Eight

The Hylorealism of Mario Bunge

The rise of what is widely known as "systems theory" (or emergent materialism) has been explicitly linked by many scholars to what Murray Bookchin (1986) described as the "modern crisis". This entailed, as noted in chapter three, the following: the concentration of economic wealth and growing social inequality; a "dialectic of violence" reflected in the disintegration of local communities, widespread genocide, and political oppression; the stockpiling of weapons of mass destruction and high levels of military expenditure; and, finally, an ecological crisis involving the pollution of the atmosphere leading to global warming, widespread deforestation, industrial forms of agriculture, the creation of toxic wastelands and a loss of species diversity (Ekins 1992; Morris 2004B: 15-17).

For Bookchin, of course, it was not simply the mechanistic worldview of Rene Descartes, and its anthropocentric ethic, that had brought about this "destruction" and this "crisis". Rather, the roots of this crisis lay firmly with an economic system—global capitalism—that saw no limit to industrial progress and no limit to growth and technology, a system that was continually "plundering the earth" in its search for profits. For Bookchin, the capitalist economy had become a "terrifying menace" to the very integrity of life on Earth, and was essentially "undoing the work of organic evolution". Industrial capitalism, he argued, was fundamentally "anti-ecological" (Bookchin 1971: 60-67, 1980: 35; Purchase 1997: 14; Kovel 2002; Morris 2012: 180-187).

For some scholars, "systems theory" and the ecological worldview that it entails are virtually equalled with Eastern mystical traditions in their varied forms. This is problematic, to say the least, if not downright obfuscating (c.f. Capra 1975, 1982: 323).

I shall specifically discuss systems theory (and emergent materialism) in the next chapter. But first, as a preface or background to this discussion, I shall focus in this chapter on the hylorealism of Mario Bunge. The chapter consists of two parts. In the first part I outline Bunge's synthesis of realism and philosophical materialism as well as his advocacy of scientific (or critical) realism; in the second part I discuss Bunge's critique of the main forms of anti-realism, namely, philosophical hermeneutics, phenomenology, social constructionism, and logical positivism .

Scientific Realism

One thing I think can be said for certain about the Argentine theoretical physicist and self-taught philosopher Mario Bunge, is that he is an articulate, trenchant, and unwavering defender of the Enlightenment tradition. Highly critical of all cultural movements (especially postmodernism) and all academic philosophy that attempts to devalue or undermine the Enlightenment, Bunge always staunchly defended the core values and tenets of this tradition. These were, namely, its realism, its philosophical materialism, its search for truth with respect to a vital emphasis on both reason and empirical science, and, finally, its democratic values and respect for the individual. As Bunge wrote: "The Enlightenment gave us most of the basic values of contemporary civilized life, such as trust in reason, a passion for free enquiry, and egalitarianism." (1999B: 142).

But for an anthropologist Bunge had his blind spots; given his strong emphasis on scientific realism he tended—like his friend Ernest Gellner—to be dismissive of phenomenology and hermeneutics. He also lacked, I think, a truly ecological sensibility. He is virtually unknown to anthropologists—even though the concept of "ontology" is paraded around at their academic conferences (it simply means the study of things)—and he has been almost totally ignored by academic philosophy. Bunge was fundamentally a philosopher and a defender of science, although he also made insightful contributions to the philosophy of the social sciences (Bunge 1996, 1998).

Mario Bunge always tended to affirm that there was a dialectical relationship between philosophy and the sciences, whether natural or social. That said, he also always repudiated the very idea of "dialectics", identifying it with Hegel's idealist metaphysics. Hegel's philosophy, he felt, was obscurantist and detrimental to scientific knowledge. Even so, he acknowledged a "watered-down" version of dialectics that emphasized variety and change (Bunge 2001B: 36-40). But Bunge constantly emphasized that philosophy and science shared many concepts and principles—the limitations of ordinary experience, the importance of the search for truth, and the linking of empirical observations with theory—concluding that "Without philosophy, science loses its depth, and without science philosophy stagnates and becomes irrelevant." (2012: 138).

It was Bunge's opinion that the influence of German idealism and the neo-romantic movement (especially with regard to Hegel, Nietzsche, and Heidegger) had led to a crisis in philosophy and the undermining of the Enlightenment within anthropology and the social sciences. This was clearly evident, he felt, in such currents of thought as critical theory, phenomenological sociology, interpretive anthropology, ethnomethodology, and postmodernism (textualism)—all of which Bunge critiqued. For what typified these neo-romantic currents, he argued, was a distrust in reason, subjective or cultural idealism, relativism, an obsession with symbols and metaphors, and a general disdain for the empirical sciences (1999B: 129-143, 2001A).

Mario Bunge was an ardent, consistent, and, at times, rather brazen advocate of philosophical materialism, which I outlined in chapter five. The materialism that he advocated—and described in detail in numerous books—was one that was realist, systematic, historical (evolutionary), and emergent.

For Bunge, the world (reality) was composed not of "facts" (as Wittgenstein [1922] alleged), nor of "events" (as Whitehead [1929] suggested), but of *concrete* material entities. Facts were simply the changing state of a concrete entity, whether facts are conceived as events, as actions, or as phenomena. For Bunge, then, the world in which we humans live is composed of concrete material entities or things, whether natural or artificial, whether living beings or inanimate objects. What is fundamental and characteristic of all concrete (material) things is that they possess a universal property, namely energy, and are thus constantly undergoing change. The world, Bunge emphasized, is not just a pile of facts: it is a *system* of concrete things because all of its components interact in complex and diverse ways with many other things. But Bunge was insistent—rightly so—that there are no properties (whether we consider energy or various qualia), no events or processes, and no relations independent of concrete material things and their changes. Disembodied relations, events, and properties are unknown; neither are there things without properties and relations. In particular, Bunge always emphasized—contra Whitehead and Deleuze—that events and processes are "what happens to, in, and among concrete systems" (Bunge 1967: 153-156, 2001B: 52-56).

Equally important, as earlier noted, is that Bunge continually affirmed that time and space only exist as *relations* between concrete things and events (2001A: 10). As we noted previously, Bunge repudiated the idea that quantum physics had "dematerialized" the world—as had been suggested by the so-called Copenhagen interpretation (Bohr 1958; Heisenberg 1958)—in several critical essays (Bunge 2012: 147). Indeed, Bunge always highlighted the fact that philosophical materialism is the *bete noire* of all religions, all spiritualist doctrines, and all forms of philosophical idealism.

Although Bunge stressed that the world is composed exclusively of concrete or material entities—together with their properties, actions, and relations—this did not imply that materialism denied the importance of ideas, or the mental activities of many forms of living beings—especially birds and mammals. But Bunge emphasized that human mental functions, whether involving perceptions, feelings, emotions, or cognition, are fundamentally brain processes that have *emerged* during the course of human social evolution. Thus, as a philosophical materialist, Bunge did not deny the existence and importance of ideas and symbolic forms, he simply denied their self-existence independent of humans (Bunge 1999A: 179).

The concept of realism, like many other complex philosophical concepts, has a diverse range of meanings. Bunge essentially recognized four distinct varieties of realism: ontological, empirical (naïve), platonic, and scientific.

Ontological realism, which Bunge fervently embraced, was outlined in chapter four. It entails the notion that material things exist in themselves, completely independent of sense experience and cognition, and that they can be known. Realism thus asserts both the autonomy and knowability of things in the material world (Bunge 1996: 353). As Bunge expressed it: "Philosophical realism, or objectivism, is the view that the external world exists independently of our sense experience, ideation and volition, and that it can be known." (2001B: 338).

Knowing the material world is attained mainly through a combination of sense experience and reason, and all factual knowledge is hypothetical and fallible. Only knowledge based on religious faith or mystical intuition, as claimed by idealist scholars, can be regarded as certain or apodictic. Ontological realism is presupposed by both our common-sense understandings of the world (in all societies) and by the empirical sciences.

Hilary Putnam's famous definition of "metaphysical realism" is quite different, and rather misleading. With regard to this perspective, he wrote that "The world consists of some fixed totality of mind-independent objects. There is exactly one true and complete description of 'the way the world is'. Truth involves some sort of correspondence relation between words or thought-signs and external things. I shall call this the externalist perspective because its favourite point of view is a God's-eye point of view." (Putnam 2002: 517).

The world does not consist of a *fixed* number of things, for it entails an evolutionary process with emergent forms; it does not imply *one* description of the world, for realism and epistemological perspectivism are not opposed concepts; and it certainly does not involve the so-called "God's eye" view of the world, which is largely an idealist fantasy. The correspondence theory of truth is, of course, a semantic issue, quite separate from realism as an ontology. As Bunge remarked after attacking realism for decades, Putnam (1994) eventually acknowledged its relevance! (1996: 354).

All knowledge presupposes a distinction between the subject, as knower, and the object of study. It further involves a complex and dynamic relationship between the human organism and the external world. The subject is neither the sole creator of knowledge (idealism) nor simply a blank slate on which the world writes its script (empiricism).

Significantly, ontological (and critical) realism retains the distinction, first broached in the seventeenth century, between the thing itself (as it exists in reality) and the thing as it is experienced by humans (appearance). This distinction is denied by empiricists (positivists), who not only view sense experience as the only source of knowledge, but in ontological terms conflate the world with our perceptual experiences of it.

Although materialism and ontological realism are distinct concepts, Bunge viewed them as intrinsically interconnected. He coined the term hylorealism to express his own approach to the understanding of the material world—both natural and social (Bunge 2006: 33).

The second form of realism is that of common-sense understanding, usually termed, rather disparagingly, as *naïve realism*. Acknowledging the existence of the world independently of humans, empirical (or naïve) realism is the view that takes our sense experiences for granted—as depicting a true *reflection* of the material world. In philosophy, this form of realism was famously expressed by Ludwig

Wittgenstein's (1922) "picture-theory" of language, namely that our ideas simply "reflect" or are an "image" of things or empirical facts.

Naïve realism was equally well depicted in Vladimir Lenin's *Materialism and Empirio-Criticism* (1972). This book stridently critiqued the phenomenalism of the positivists Richard Avenarius and Ernst Mach, and their Russian disciples. As a physicist and philosopher of science, Mach had contended that all things were simply a complex of sensations, thus implying that all science could be reduced to psychology. In Neo-Kantian fashion, Mach dismissed any reference to a natural world prior to humans, or to "things-in-themselves", as pure metaphysics (not science), even refusing to acknowledge the existence of atoms. Lenin argued that Mach and Avenarius were simply following in the footsteps of the Irish cleric George Berkeley, a subjective idealist who even denied the reality of the material world.

Lenin, concerned to defend what he felt was the dialectical materialism of Marx and Engels—along with that of Feuerbach and Joseph Dietzgen—strongly argued that the material world existed independently of human consciousness and sense-experience, and that our sensations were "images" of an objectively real external world (realism). He acknowledged that such scientific materialism had been dubbed "naïve realism", thus implying that it was the basis of common-sense understanding (1972: 426). Lenin sums up the contrasting perspectives of Marxism and the Neo-Kantian empirio-criticism of Mach and Avenarius as follows, citing Berkeley: "Materialism is the recognition of objects in themselves, as outside the mind; ideas and sensations are copies or images of those objects: The opposite doctrine (idealism) claim that objects do not exist 'without the mind'; objects are 'combinations of sensations'" (1972: 14).

Inadvertently, while criticizing the subjective idealism or phenomenalism of Mach and Avenarius, Lenin embraced a form of naïve or common-sense realism. This form of realism, like that of the classical psychological empiricists (Locke and Hume), tends to conflate ideas (cognition) with our sense impressions, and holds that knowledge (or language, as with the early Wittgenstein) is simply a direct "copy", "image", or a "mirror of nature" (Rorty 1980)—simply a *reflective* representation of the material world. There is then, for the empirical (naïve) realist, an isomorphic relationship between knowledge (concepts), and reality (things) (Greenwood 1991: 82; Bunge 1996: 354). This may have a certain survival value for humans, for the very idea of factual or objective truths—such that crops need water to flourish—presupposes a *realist* perspective.

But naïve realism, in privileging sense impressions (phenomena), is, Bunge argued, a very limited perspective, for it is unable to give any account of unobservables—imperceptible things such as atoms, molecules, dinosaurs, and social systems—or of imperceptible processes in both the natural world and human social life. For Bunge naïve realism (empiricism) was thus uncritical, and particularly prone to sensuous delusions and self-deception. As he put it, this form of realism was a "sitting duck for sceptics and idealists" (1996: 354-355, 2001B: 339-340).

Yet, for Bunge, a rejection of empiricism or native realism did not entail a rejection of ontological realism, nor a nihilistic attitude towards the concept of "representation" and thus a lapse into subjectivism. It entailed, rather, the development of an alternative critical or scientific realism (see below).

The third form of realism, Platonic realism, is, as I outlined in chapter two, essentially a spiritualist metaphysic, a form of objective *idealism*. Knowing, for Plato, consisted of the intellectual intuition (*theoria*) of eternal forms, and he tended to have a very disparaging attitude towards the body, the senses, the emotions, and empirical knowledge. Concrete material things, for Plato, were simply the imperfect copies of ideas or forms (*eidos*). These forms had an autonomous existence in some divine realm. Bunge's hylorealism is the exact antithesis—not simply a reversal—of Platonism, emphasizing that reality is composed entirely of concrete material things, and that ideas, rather than being self-existing (as with Plato), are processes occurring within the brains of some animals (Bunge 1996: 43).

The final form of realism is the one strongly advocated by Mario Bunge, namely critical or *scientific realism*. It acknowledges the ontological thesis that there are material things that exist "in-themselves", independent of human cognition, and that these things are knowable not by some mystical or philosophical intuition, but by a combination of empirical observations (experience) and reason (specifically, theorizing).

Bunge emphasizes the limitations of phenomenal experience. A phenomenon is what appears to a person in subjective experience. Bunge argued that such phenomenal knowledge, based largely on *perceptual* experience, was limited, rather shallow, and "skin deep". For real knowledge of material things and processes is neither direct nor a pictorial representation (Bunge 2001B: 27-28).

As Bunge writes with respect to critical reason: "Scientific research seeks reality beneath appearances, because the latter is superficial and subject-centred, whereas scientific knowledge is profound and objective...the search for pattern is what scientific research is all about. This research takes us beyond perception into conception, and particularly into theory...explaining appearances in terms of hypotheses that posit imperceptible things and processes." He goes on to state: "The scientific realist does not confine himself to appearances, but he does not write them off either. On the contrary, he often starts from them and attempts to explain them." (1996: 42-43). Needless to say, even in everyday life, people seek the reality behind appearances—a distinction some philosophers fail to grasp—and attempt to explain events—happenings—in both the material and social worlds.

I shall return to issues relating to realism and scientific explanations later in the manifesto, but we may turn now to Bunge's critique of the various forms of anti-realism, which I broached in chapter four, and—in the next chapter—to his advocacy of emergent materialism as a systemic worldview.

A Critique of Anti-Realism

Philosophical idealism (or spiritualism) is a doctrine that takes many different forms. To give a rough sample, it is the view that only ideas have reality; or, that the natural world does not exist independently of ideas or mental representatives; or, more prosaically, that ideas have primacy over material relations in our understanding of the world.

Mario Bunge was of the opinion that the subjective idealism of Immanuel Kant had been a decisive influence not only in the development of orthodox Neo-Kantianism—well represented by Wilhelm Dilthey's philosophical anthropology (see Morris 2014A: 406-415)—but also on several other currents of thought that had emerged during the twentieth century. These include, in particular, philosophical hermeneutics, phenomenology, social constructionism (otherwise known as postmodernism), and logical positivism. What they all had in common, according to Bunge, was their opposition to both ontological and critical realism in favour of philosophical subjectivism. This is the view that the world, far from existing independently of the human subject, was, in fact, the "creation of the knowing subject" (Bunge 2001B: 323). Each of these four philosophical currents of thought thus rejected hylorealism—the thesis that the world consists of material entities that exist "in-themselves", independent of the knower, and that scientific (or critical) reason enables us to understand and explain many aspects of the real world (Bunge 2006: 57).

I will briefly outline Bunge's critique of each of these currents of thought in turn. I will make specific references to the social sciences, which were of particular interest to Bunge as a philosopher of science.

Philosophical Hermeneutics

Hermeneutics is simply the art or science of interpretation—specifically understanding the meaning or significance of a text or human action. The Greek word *hermeneuein* means to interpret or translate. The Latin translation of the Greek world is *interpretio*, so in an important sense hermeneutics and interpretation refer to the same intellectual activity (Schmidt 2006: 6). What was significant, and troubling, for Bunge was that hermeneutic scholars—Ernst Cassirer (1944), Clifford Geertz (1973), and Hans-Georg Gadamer (1976), being among the key figures in this tradition—claimed that all social facts are cultural or symbolic—or even spiritual—and must accordingly be interpreted as if they were a text. For Bunge this was a thoroughly idealistic and limiting perspective. For social facts—such as funerals, riots, productive work voting, or organizing a jumble sale—as well as more enduring social institutions—like the state—cannot be viewed simply as a collection or "web" of symbols, to be interpreted as if they were texts (textualism). To the contrary, such social phenomena, Bunge stressed, have to be described and explained, even if the interpretation of meanings is intrinsic to social analysis. As Bunge writes: "Since humans are among other things, symbol-making animals, who think and inter-act with the help of symbols, it would be foolish of a social scientist to disregard symbols altogether. But to regard individuals and societies as texts is an idealist extravagance." (Bunge 1996: 293; Gellner 1985).

Hermeneutics, of course, along with interpretive anthropology was an off-spring of Neo-Kantian philosophy. Firmly rejecting materialism, this philosophy advocated a radical cleavage between the material (nature) and the spiritual (culture) realms of existence. This also implied a radical distinction between the natural sciences—concerned with the formulation of causal laws and the *explanation* of natural phenomena (thus nomothetic)— and the cultural sciences—concerned with history as an idiographic study and with the *understanding* of human social life and culture (Morris 2014A: 406).

Bunge, of course, as an emergent materialist, firmly repudiated the many dualisms (culture/nature, idiographic/nomothetic) that were inherent in Neo-Kantian philosophy (Bunge 1996: 290-293). Bunge, therefore, always emphasized the limitations of both philosophical hermeneutics (Dilthey, Gadamer) and interpretive anthropology (Geertz). He highlighted, for example, that they tended to shun the methods of the empirical sciences, and thus offered little in the way of *explaining* social phenomena. Furthermore, Bunge argued they were singularly unaware of the complexity and polysemy of the concept of "meaning", and, in focussing on daily life and symbolic forms, failed to engage with macro-social events and processes, specifically the genesis and changing structure of social systems (on the limitations, for example, of Geertz's understanding of Moroccan politics see Morris 2006: 105-110). Bunge thus concluded that to understand the complexity of human social life we must go beyond intuitive understanding (*verstehen*) and engage in the scientific study of social reality (Bunge 2011A: 122-136).

Phenomenology

Phenomenology simply means the science (*logos*) of phenomena—while the term phenomenon, derived from the Greeks, refers to something that "appears" to human consciousness.

Phenomenology is indubitably associated with the German philosopher Edmund Husserl, along with his student Martin Heidegger, whose own existential phenomenology essentially combined Husserl's phenomenology with Nietzsche's nihilistic philosophy and Dilthey's historicism. I have no wish here to become embroiled in Husserl's abstruse phenomenology, which was expressed in a prose style that was dense, convoluted, and austere, and which, according to Bunge, often degenerated into scholastic "gibberish" (2003: 55). I have myself attempted to portray a more sympathetic account of Husserl's phenomenology (2014A: 504-525), for Husserl seems to have spent almost every day of his life writing down his reflections, and meditations, on the nature of human consciousness (Moran 2005: 5-42). He was indeed an ultra-rationalist, advocating an a priori rationalism even more extreme than that of Descartes (Lichtheim 1972: 224). In fact, Husserl described his own philosophy as a form of "transcendental idealism", and the world itself as an "infinite idea" (1931: 86-97).

Phenomenology, as Husserl envisaged it, implied the advocacy of an eidetic science, which, through transcendental reduction or *epoche*—suspending or "bracketing" the world of everyday experience—attempted to simply describe (not explain) the basic structures or meaning of human consciousness and understanding, largely from a first-person or subjective point of view.

Although never doubting the importance and significance of the *natural* sciences, which he considered one of the highlights of the human spirit, Husserl essentially put up a "picket fence"—as Stephen Jay Gould described it (1984: 241)—around human consciousness and human social life and culture, making them the exclusive preserve of philosophy (as he conceived it) and the humanities. Thus, taking Neo-Kantianism to an extreme, Husserl distanced himself from the empirical rationalism of the Enlightenment, and was not only "anti-historical" (Arendt 1978: 9), but vehemently opposed to empirical psychology and the social sciences. Both, for Husserl, implied the "naturalization of the spirit" (1970: 290-292), and were therefore interpreted as a reductive form of naturalism. As he put it, an objective science of spirit (human consciousness and culture) "has never existed and will never exist" (1970: 297). Husserl's phenomenology is thus the exact antithesis of positivistic sociology, sociobiology, and emergent materialism.

Given Husserl's explicit and rather arrant rejection of the psychological and social sciences, it is hardly surprising that Mario Bunge was highly critical of Husserl's phenomenology, along with its offspring, phenomenological sociology (Schutz 1967). He rightly described it as a form of "subjective idealism", and summed up the contribution of phenomenology to social studies with the following dismissive note: "impenetrable, pretentious jargon, lack of rigour, subjectivism, contempt for the empirical research, and exclusive interest in the minutiae of everyday life at the expense of their social and historical context" (Bunge 1996: 295). The Argentine scholar thus concluded that the a priori theorizing and subjectivism of Husserl's phenomenology entailed a "drastic retreat from reality" (2006: 64-65). How could anyone, he wrote, think of phenomenology except as a "wild fantasy", a metaphysic unable to "shed any light on anything except the decadence of German philosophy" (1999B: 211)?

What particularly concerned Bunge was that both hermeneutics and phenomenology, and their exclusive forms on *lebenswelt* (the life-world), tended to explicitly disregard such macrosocial processes as social conflicts, economic production and cycles, and wars (2012: 9).

Social Constructionism

In the opening chapter of this manifesto, I discussed the tendency of postmodern anthropologists to adopt a Neo-Kantian (idealist) perspective, and, as social constructionists, to virtually deny the reality of the material world. In chapter four, I offered further critical reflections on this form of anti-realism, which became quite fashionable among literary theorists and many social scientists towards the end of the twentieth century.

Bunge, however, always made a clear distinction between *epistemological* constructionism, which was almost universally acknowledged and practised among scholars, and *ontological* constructionism, which was the purview of subjective idealists, postmodernists, literary theorists, and many sociologists who were involved in the social study of science (e.g., Latour and Woolgar 1979; Knorr-Cetina and Mulkay 1983). All these scholars seem to deny that the material world—the things-in-themselves and the facts about them—exist independently of human cognition.

Apart from radical empiricists, for whom ideas emerge directly from our perceptions of the world—the things themselves, though existing autonomously, being considered simply as a "collection" of sensations (the phenomenalism of Ernst Mach)—almost all scholars affirm that ideas (concepts) are social constructs. Both human ideas and human knowledge are social constructions that emerge from our interactions with the world, both natural and social. As John Dewey (1925) emphasized, human knowledge is *produced,* socially constructed. Epistemological constructionism is therefore a perfectly valid and essential pursuit, and, as Bunge stressed, scholars such as Aristotle, Engels, Einstein, Piaget, and Popper all held that our concepts and scientific theories were human constructions (1996: 295). Not only scientific theories but all cultural worldviews and symbolic forms (representations) are social constructions, generated within particular historical societies.

But *ontological* constructionism, Bunge argued, was a very different thing, for it was a form of idealism that denied the material world its reality—or in the words of Arthur Schopenhauer (1819), "The world is my idea (representation)." (Bunge 1996: 296).

Bunge defined ontological constructionism as "the view that the world is a human construction; that there are no things in themselves but only things for us (humans). According to this view, nature has no independent existence. This thesis is in violent conflict with everything we know about the world before the emergence of human beings." (Bunge 1999A: 48).

Social constructionism, as an ontological thesis, is a combination of anti-realism and collectivism (holism); it is essentially a sociological or cultural version of Neo-Kantian idealism. As a philosophical paradigm, social constructionism, as Bunge continually insisted, tended to conflate or confuse reality with our representations of it: the world with cultural worldviews, the explored with the explorer, the territory with its maps (1996: 336).

In an introduction to "social constructionism" the psychologist Vivien Burr not only—mistakenly—conflates realism and empiricism (naïve realism), and—again mistakenly—assumes that our knowledge of the world comes entirely from other humans, but writes that "when people talk to each other, the world gets constructed" (1995: 6-8). Similar sentiments have been expressed by many other scholars, as discussed earlier. A fish is only a fish, wrote one, if it is "socially classified as one" (Tester 1991: 46). Another anthropologist writes that "tigers are socially constructed" (Jalais 2010: 196), meaning simply that people have multiple and contrasting understandings of the living animal, which exists, of course, in its own right.

This kind of cultural (or linguistic) idealism was essentially derived from the oracular statements expressed by three iconic figures of postmodernism; namely, "there are no facts only interpretations" (Nietzsche 1967: 267); "there is nothing beyond the text" (Derrida 1976: 158); and "discourses...form the objects of which they speak" (Foucault 1972: 49). Whether these scholars meant such statements to be taken literally is debatable!

Given the brute reality of the material world existing quite independently of humans, recognized not only by scientific realists like Bunge but by *all* humans in their everyday life (including the social constructionists themselves!), such onto-logical constructionists engage in rather convoluted theoretical discussions that attempt to reconcile the cultural idealism they stridently proclaim with their own common-sense understandings of the world! Only a cultural idealist would pose the question: "Is there a real world outside discourse?" (Burr 1995: 81). Apart from idealist scholars, whether of subjective or cultural (social) persuasion, most people would respond to this question in the affirmative. As a materialist and on-tological realist, Bunge is *emphatic* that there is, and that it can be known.

But although Bunge is critical of the cultural (or linguistic) idealism that is associated with Derrida, Foucault, and their postmodern acolytes, as a philosopher of science he reserved his most critical ire for a coterie of scholars whom he grouped together as advocates of the "constructivist-relativist sociology of sci-ence". Inspired by the writings of Thomas Kuhn (1962) and Paul Feyerabend (1975), these scholars (for example, Bloor 1976; Latour and Woolgar 1979; Knorr-Cetina and Mulkay 1983), according to Bunge, adopt an extreme form of ontolo-gical constructivism, holding that not only are theories and experiments social constructs, but that reality itself—along with scientific facts—are also social con-structions. Collapsing the distinction between the material world and our repres-entations of it, these sociologists of science also proclaim "the abolition of the fact/theory distinction" (Barnes 1983: 21). But, as Bunge affirms, if facts (things and events) and theories (conceptual frameworks) are considered the same, "no fact could be used to test a theory, and no theory could be used to guide the search for new facts" (1999B: 174).

Moreover, if there is no independent reality and the world is deemed to be a social construction, and if facts are only statements of a certain kind, there can, Bunge argues, be no objective truth. Sociologists of science like Latour and Bloor are thus not only ontological constructionists but embrace a form of epi-stemological or cultural relativism. The concept of (empirical) truth as the cor-respondence of "ideas with facts" thus, for these scholars, ceases to make sense. But for Bunge, to the contrary, science *is* about the "quest for truth" with respect to both the natural world and human social life and culture (for an extensive critique of the ontological constructionism of the aforementioned sociologists of science, see Bunge 1999B: 173-207).

Logical Positivism

To many cultural anthropologists, any scholar who advocates a *scientific* approach to human social life and culture is invariably (and misleadingly!) described as a "positivist". Such anthropologists, like Lukacs (1971), therefore tend to equate the natural and social sciences with positivism, and seem to denigrate both. Indeed, as Bunge remarks, "positivism-bashing" is highly fashionable among contemporary scholars—whether ethnomethodologists, phenomenologists, feminist theorists, or existentialists. Many of these same scholars, somewhat ironically, behave like positivists—as radical empiricists—while critiquing and denouncing positivism! This is a rather devious way, Bunge contended, of attacking any *scientific* approach to human social life and culture (1999B: 213). It is therefore important to recognize that neither materialism nor the social sciences can be equated with positivism.

Positivism, as a philosophical worldview, has its roots in the early empiricist philosophy of Francis Bacon, John Locke, and David Hume. It flourished in the nineteenth century, being particularly associated with Auguste Comte, Herbert Spencer, and John Stuart Mill. Generally regarded as a materialist philosophy that was both anti-clerical and politically radical, positivism came into its own in the 1930s. For during this decade, a group of scientists and philosophers widely known as the Vienna Circle (1926-1936) established the well-known philosophical outlook named logical positivism. Taking its inspiration especially from the writings of Ernst Mach, logical positivism combined an emphasis on logical analysis, mathematics, and the exact sciences (especially modern physics); a strong aversion to metaphysical thinking, particularly that associated with the obscurantist philosopher Martin Heidegger; and, finally, a radical form of empiricism. Logical positivism, though anti-metaphysics, expressed, of course, its own metaphysics, that of radical empiricism. The logical positivist perspective was particularly associated with such scholars as Moritz Schlick, Otto Neurath, Rudolf Carnap, and A.J. Ayer. Indeed, Ayer's study *Language, Truth and Logic* (1936) did much to popularize logical positivism as a philosophical movement.

Unlike those scholars who engage in "positivism-bashing", Bunge recognized the virtues of logical positivism, namely its conceptual clarity, its critique of philosophical obscurantism, its stress on the importance of empirical testing, and its affirmation of the scientific method as a way of understanding the material world (1996: 317).

But Bunge also highlighted the philosophical limitations of logical positivism and offered several important and telling criticisms.

The first concerned its embrace of radical empiricism—its phenomenalist metaphysics. Derived from Mach, and stressed by Carnap, this metaphysic in essence rejected both materialism and critical realism, both of which were dismissed by the logical positivists as "metaphysical nonsense". Phenomenalism, the notion embraced by the logical positivists that everything is merely a phenomenon—material things being merely a collection of sense impressions (or data)—is in fact close to subjective idealism in denying the reality of things "in themselves", that is, independent of our perceiving them (Bunge 1996: 283).

Bunge rejected such phenomenalism as not only contrary to modern scientific understanding but also to ordinary knowledge. For phenomena or appearances are only the starting point for any inquiry, and even in everyday social life we seek reality—explanations—behind the appearances (Bunge 1999A: 208). Phenomenalism, in opposing realism, was for Bunge thoroughly anthropocentric.

Always making a clear distinction between empiricism and realism, Bunge thus wrote: "Whereas realists attempt to account for reality, empiricists—from Ptolemy to Hume, Kant, Comte, Mach and Carnap—stick to phenomena or appearances. Thus they are anthropocentric, since there can be no appearances without subjects" (2012: 20). Thus empiricism was viewed by Bunge as an obstacle to scientific research in being subject-centred. In contrast, Bunge advocated scientific realism, the attempt to understand the material world and human social life as objectively as possible.

Secondly, and linked with this, logical positivism emphasized that science deals only with empirical facts and that scientific knowledge consists essentially of analytic inductive generalizations based on observations and sense impressions. It followed Hume in interpreting causality subjectively as the regular conjunction or succession of events. In the adoption of a strictly empiricist theory of knowledge—the view that all knowledge is based on sense impressions—he advocated a form of descriptivism. Logical positivism thus rejected explanations in terms of hidden causes, specifically causal mechanisms, and was concerned only with observed facts, with empirical generalizations, and with input-output analysis. For Bunge, this strictly curtailed scientific research (1999B: 29).

As empiricists, logical positivists were committed to what came to be known as the verification principle. According to this principle, constructs are meaningful only if they can be verified by sensory experience. This, of course, ruled out the existence of such unobservables as atoms. Other positivists who came to be described as logical empiricists—such as Otto Neurath—therefore embraced a form of reductive materialism or physicalism. As an emergent materialist, for Bunge this was equally untenable (Greenwood 1991: 83-84; Bunge 1996: 129, 1999B: 28-30).

Thirdly, logical positivism stressed a fundamental dichotomy between facts and value statements. Judgements of value or morality were held to have no empirical content of the sort that renders them accessible to tests of validity. It thus favoured an emotive theory of ethics, viewing philosophy as simply the "handmaiden" of the natural sciences. Bunge firmly rejected the axiology (value theory) of the logical positivists, emphasizing that social actions, human values, and morals (ethics) should be firmly rooted in a realist and materialist philosophy (Bunge 2001A: 205).

Finally, like Edmond Wilson in a later decade, logical positivism emphasized the essential unity of the natural and social sciences, suggesting that they share a common logical and methodological foundation. The procedures of the natural sciences are, therefore, applicable to all spheres of social life and culture. As Rudolf Carnap put it, "There is no question whose answer is in principle unattainable by science." (1967: 290). The unity of the sciences as envisaged by logical positivism tended to give epistemic privilege to mathematics and physics. Denying the autonomy of the biological and social sciences, invariably entailed, according to Bunge, a form of reductive materialism or physicalism. In contrast, Bunge advocated systems theory and a form of emergent materialism.

Although there are few philosophers around who would now describe themselves as logical positivists, positivist tendencies are still evident in the psychological and social sciences—as Bunge continually stressed. Such tendencies can be seen, for example, among ethnomethodologists, psychological behaviourists, phenomenological sociologists, network theorists, and sociobiologists (on positivism see Outhwaite 1987: 5-11; Giddens 1996: 151-161; Morris 2014A: 199-200; Bunge 1999B: 28-30).

After this introductory account of hylorealism, I shall turn now to the third form of philosophical materialism—that of emergent materialism, a theory particularly well-expressed in the systems theory of Mario Bunge.

Chapter Nine

Emergent Materialism

In this chapter's discussion of emergent materialism (or systems theory) some preliminary reflections seem perhaps necessary.

The Turning Point

There has been a widespread and lamentable tendency among many scholars, especially eco-spiritualists, to equate Western philosophy, even Western culture, with the mechanistic philosophy of Newton and Rene Descartes. This philosophy or paradigm is described by such scholars in terms of the following three tenets: a dualistic metaphysic that radically separates humans from nature, implying a worldview that envisages nature (the earth) as lifeless, mechanical, and as consisting of isolated material things bereft of relationships; an epistemology that is atomistic, reductionist, and anthropocentric; and, finally, as entailing an ultra-rationalistic ethic that sanctifies the technological domination of nature by humans. It is as if, as St. Augustine of Hippo (1991) alleged, the deity had created the world solely for human use—though, not, of course, to be aesthetically enjoyed!

According to these same scholars, specifically the spiritualist systems theorists and deep ecologists (Bateson 1979; Capra 1982), during the 1960s a "turning point" occurred within Western culture, with the rise of a "counter-culture". There was, according to Fritjof Capra, a "paradigm shift" (1982: xviii). Taking on the mantle of "innovators" or even as cultural "visionaries", such scholars affirmed—contrary to Cartesian philosophy—that humans are indeed an intrinsic part of nature, that all phenomena are interdependent and interconnected, and that life and consciousness cannot be reduced to physical matter. They described this "new paradigm" as a new vision of reality, one that was holistic and ecological. However, these advocates of deep ecology and the systems view of life tended—rather misleadingly!—to equate a naturalistic, ecological worldview with the varied forms of religious mysticism found throughout human history (Capra 1982: 323; Devall and Sessions 1985).

Two points may be made regarding this "turning point" scenario.

Firstly, it is important to recognize that Descartes, like most Western philosophers, was not a materialist but was instead an essentially religious thinker, a subjective idealist. To equate Cartesian philosophy with either philosophical materialism or ontological realism is simply obfuscating. Indeed, to suggest that ontological realism—the notion that nature is indeed distinct from humans—entails an ethic of domination towards the natural world (Kumar 2016: 44) is equally obfuscating. That nature has an integrity and is independent of humans does not imply that humans are separate from nature. Still less does it warrant the abusive exploitation of nature—quite the contrary.

Secondly, we need to recognize that Cartesian mechanistic philosophy has indeed been a dominant influence on Western intellectual culture, given that it is intrinsically linked to the ethos of industrial capitalism. This is something I emphasized long ago (Morris 1981), and Capra insightfully explores in relation to biomedicine, economics, and academic psychology. He emphasizes that the mechanistic Cartesian worldview has powerfully influenced both the empirical sciences and Western culture more generally (Capra 1982: 118-247).

But it is also important to recognize that critiques of the Cartesian worldview, its dualism and its anthropocentrism, did not suddenly emerge as a "turning point" in the 1960s. These critiques instead go back to the eighteenth-century Enlightenment (which I discussed in chapter three). Indeed, the development of an ecological worldview—which combines evolutionary scientific naturalism and an aesthetic appreciation of the natural world, and emphasizes the organic (not spiritual) unity of humans and nature in a complex web of life—was developed in the nineteenth century, and by many scholars. For a naturalistic, ecological worldview was intrinsic to the writings of such nineteenth-century naturalists and radical scholars as Alexander von Humboldt, Charles Darwin, Karl Marx, Ernst Haeckel, Peter Kropotkin, and Elisée Reclus, who hardly get a mention in the writings of Fritjof Capra and the deep ecologists. An ecological sensibility was also clearly expressed at the turn of the century by many literary naturalists; for example, John Muir, Richard Jefferies, W.H. Hudson, John Burroughs, and Ernst Thompson Seton (see, for example, Morris 1981, 2004B; Foster 2000; Clark and Martin 2004; R.J. Richards 2008; Wulf 2015).

But this chapter focusses not on the spiritual ecology of such system theorists as Capra, but on the emergent materialism of Mario Bunge. For Bunge's writings are lucid, coherent, engaging, and substantive, in marked contrast to the dense, obscure prose of the "continental" philosophers who have recently—almost a hundred years after Roy Wood Sellars!—discovered for themselves the relevance of a materialist and realist ontology (Meillassoux 2008; Gratton 2014). As Bunge put it: "There is nothing like forgetting the past to spew apparently new philosophies." (2010: viii).

Systemism

Systems form relationships and patterns with other systems. Interaction between a diversity of systems, combined with unique environments and histories, creates novelty, invention, adaptation and evolution. And there you have it! The story of life, the universe and everything within in! (Purchase 1995: 1)

Systemism, for Mario Bunge, was a worldview, a cosmology, a way of thinking about the world. He recognized that many of the key concepts of such a systemic philosophy were derived from evolutionary biology. These were, specifically, the concepts of evolution, levels of organization, and emergence (Bunge 2003: xiii).

Like Humboldt, Bunge acknowledged that the world itself, the cosmos, was a system in that it was comprised or characterized as a universal interconnectedness of things that were in a constant state of flux. Bunge's ontological thesis therefore stressed that every *thing* or *object*—whether concrete (material), conceptual, or symbolic—was either a system or a component in one or more systems. He thus strongly emphasized the "inter-connections among things" (1996: 195, 2001A: 40).

Unlike the deep ecologists, Bunge recognized that systemic philosophy and the concept of emergence had a long history going back to the Enlightenment. He specifically mentioned Paul-Henry D'Holbach's well known *System of Nature* (1770), which combined a strident mechanical materialism with a systemic philosophy. Bunge went on to note that Systemism was every bit as unpopular among contemporary academic philosophers, who, like Plato, equate reality with Chaos, as philosophical materialism and the Enlightenment (Bunge 2003: 41).

But Bunge also highlights the fact that many insightful scholars since the Enlightenment have expressed a systemic philosophy, and recognizes the importance of emergence as a property of all complex systems. John Stuart Mill (1843)—and many other scholars—for example, recognized that water was a "liquid emergent", a compound derived from a combination of the elements oxygen and hydrogen, and that it had properties very different from those of its components (Wheeler 1928: 15; Bunge 2003: 16). Likewise, the polymath George Henry Lewes (1875), a keen and insightful interpreter of both Aristotle and Goethe, clearly recognized that life, mind (psyche), and human social life and culture were all "emergent" phenomena.

During the 1920s, Roy Wood Sellars (1922), Conwy Lloyd Morgan (1923), and William Morton Wheeler (1928) all published studies affirming the importance of what was then described as "emergent evolution". Lloyd Morgan, for example, suggested that humans lived in a world in which there seemed to be an orderly sequence of events, reflected in three evolutionary "levels", namely those of matter, life, and mind. Although he recognized that "relatedness is an essential feature of reality" (1923: 178), Morgan, unlike Lewes, lacked a social perspective. Moreover, as a spiritualist, Morgan conceived of God as the ultimate "creative source" of evolution (Wheeler 1928: 41; Morgan 1923: 89).

In contrast to Lloyd Morgan's theological approach to "emergent evolution", Roy Wood Sellars advocated, in systemic fashion, a theory of "evolutionary naturalism". Sellars' theory essentially combines a dynamic, non-reductive form of naturalism (materialism) with critical realism. As he wrote: "Knowledge is a human affair, even though that which is known is distinct from the knower. But man is a part of nature." (1922: 2). He thus explicitly recognized the paradox of human life.

In advocating evolutionary naturalism as a worldview, Sellars made some telling critiques of Platonism, the subjective idealism of Berkeley and Kant, and the radical empiricism (positivism) of Mach and Avenarius. In defending critical realism, Sellars upheld the principle that "The objects of knowledge do not depend for either their being or nature upon our knowledge of them.... Being is one thing, and knowledge is quite another." (1922: 22). In so doing, he also made important criticisms of both naïve realism and its correlate, the "representational" or "copy-theory" of perception (knowledge)—associated, in particular, with the realism of John Locke (1922: 29). In later years Sellars became a strong advocate of humanism as a religion (Warren 1975).

Bunge warmly acknowledged the seminal efforts of such early scholars as Lewes, Sellars, and Norbert Wiener—a pioneer in the development of cybernetic systems theory (1948)—in the development of systems theory and the concept of emergent evolution (Bunge 2003: 16).

What is significant with regards to Bunge's systemic philosophy is that he always made a clear distinction between the realm of ideas (concepts, maps, theories, the products of the human imagination), and the realm of concrete, material things. There are thus two kinds of existence: real (or material) and ideal (or imaginary) (2012: 178). This did not imply a radical dichotomy, for Bunge advocated a form of emergent materialism, stressing that ideas had no existence independent of living organisms, while symbolic forms were simply conceived as the products of humans living together in social groups (2012: 19).

This led Bunge to make a clear distinction between two kinds of systems. On the one hand, there are conceptual systems (such as worldviews, scientific theories, legal codes, and symbolic forms); on the other hand, there are concrete or material systems, which may be natural or social and can consist of human artefacts or cultural environments (like meadows).

Equally important, Bunge made a distinction between two ways in which material things may conjoin to form a unity or whole. This can occur either by association, juxtaposition, or accretion, so as to form *aggregates*; or by composition, or binding relations (organization), so as to form *systems*. Some examples of aggregates include the accretion of dust particles, sand dunes, garbage heaps, clouds in the sky, a swarm of locusts, or a loose crowd of people. Aggregates are characterized by a modular structure, or even a complete lack of structure, as well as a low degree of cohesion. Systems, in contrast, are not just a random collection of parts, but have an integral structure constituted of strong, usually enduring bonds (relations). Systems are, therefore, more stable, more cohesive, and more enduring than mere aggregates. Typically, systems as complex wholes possess properties that their component parts lack. It is these global properties that are said to be *emergent* (Bunge 2003: 10-12).

As Bunge writes, "When two or more things get together by interacting strongly in a specific way, they constitute a system. This is a complex thing possessing a definite structure. Atomic nuclei, atoms, molecules, crystals, organelles, cells, organs, multicellular organisms, biopopulations, ecosystems, human families, business enterprises, and other organizations are systems. They may all be said to emerge through combination or self-organization rather than aggregation"

(2001B: 169, 2003: 27). The bonds within a system, comprising its structure (or internal relations) are always stronger than its relations with the (external) environment—by definition!

Material systems behave as a whole and have an integral unity or identity. They can also be characterized, according to Bunge, by four aspects or dimensions, which he frequently described as the CESM model. These are its *composition*, the collections of parts (things) that comprise a system; the *environment* of a system, that is, its relationship to an external world of things that affect or are affected by the system; the *structure* of the system, the relationships between the various parts or components of the system; and, finally, the system's *mechanisms*, the various morphogenetic processes that make a system "tick" or function (Bunge 1999A: 282, 2003: 35).

What must be recognized, of course, is that mechanisms only exist in concrete, material systems (such as atomic nuclei, cells, organisms, or social systems) and that mechanisms are not specifically mechanical. To the contrary, they refer to relations and processes within a system that bring about its emergence (or becoming) or lead to changes (or stability) with regard to the system as a whole. Cell division, gene mutation, and natural selection, are, for example, mechanisms with respect to biological systems. Trade, work, co-operation, and competition are all mechanisms within social systems (Bunge 2003: 21). Mechanisms are processes, and processes are what happens when material things dynamically interact.

Equally important, the CESM model of a given system says nothing about its history, and the history of a system is particularly significant with respect to biosystems—living things—and social systems (Mahner and Bunge 1997: 27).

A systemic approach to the understanding of both the world of nature and human social life was linked by Bunge to a materialist ontology. Systemism therefore entailed both a dynamist conception of matter—acknowledging that all material things (systems) were constantly in flux—and an evolutionary perspective. Thus Bunge not only distinguished between two levels of organization within a complex system, the micro- and macro- levels, but affirmed that the world itself has a "multi-level" structure. Bunge essentially recognized four levels of organization, namely the physical, chemical, biological, and social levels. He emphasized that each level of organization and complexity had its own specific properties and laws. He also stressed that no level was entirely independent of adjoining levels (2001B: 31).

Bunge recognized that further subdivisions could be made within each of these levels. For example, the biological level of organization could be subdivided into levels relating to cell, organ, organism, biopopulation, and ecosystem (2001B: 81). But, interestingly, Bunge's pluralist ontology did not acknowledge a specific *psychological* level of organization. For although he asserted that the organism endowed with mental abilities constituted a specific level of their own, he did not consider that "minds constitute a level of their own—and this is simply because there are no disembodied minds" (2001B: 89). Bunge was against the whole idea of reifying the mind in the fashion of idealist philosophers.

But Bunge further suggested that technical levels (relating to artefacts) and semiotic levels (relating to symbolic systems or culture) could be recognized within the social level of organization (2001A: 75). Such a multilevel ontology, with its emphasis on emergence and self-organization, was, Bunge recognized, "a sort of generalization of the theory of evolution" (2001B: 81).

An emphasis on the importance of recognizing that the world consists of multiple levels of organization—in marked contrast to the flat ontology of the network theorists (e.g., Latour, Ingold)—is well expressed in the school of thought known as critical realism. It is a research strategy particularly associated with the (early) seminal writings of Roy Bhaskar. Like Bunge, Bhaskar and his associates were highly critical of the anti-realism of contemporary thought—whether expressed as Neo-Kantian idealism, classical empiricism, logical positivism, or the postmodernism of Richard Rorty (Bhaskar 1991; Collier 1994).

The critical realists recognized three levels of existence: the *real* world of material things, with their causal powers and generative mechanisms; the *actual* realm of events; and, finally, the *empirical* realm, the world as experienced, recognizing that not all events are in fact experienced by humans (Bhaskar 1975: 13). In contrast to both Kantian transcendental idealism and empiricism, the critical (or transcendental) realists advocated "deep realism". They acknowledged that material things have the powers that they do because of their structures. Like Bunge, they therefore stressed that to understand phenomena (or events) it is necessary to investigate the mechanisms (or processes) that are involved in their generation. As Bhaskar writes: "The real basis of causal laws, are provided by the generative mechanisms of nature. Such generative mechanisms are... nothing other than the ways of acting of things". The idea of an empirical world, Bhaskar suggests, is therefore a "category mistake" (Bhaskar 1975: 14-16).

The critical realists also emphasized, like Bunge, that "nature is stratified"—the physical, chemical, organic, and social (human) levels relating, Bhaskar insisted, to distinct kinds of mechanisms, as well as stressing the salience of the concept of emergence (Bhaskar 1975: 119; Collier 1994: 107-115; on critical realism see Archer et al 1998).

But let us return to Bunge's systemic philosophy. In emphasizing the centrality of material systems, Bunge denied the independent existence of properties and relations, while firmly stressing that all material things have properties and relations. What his systemic approach specifically highlighted was the concept of emergence, not only in the sense that all complex systems are endowed with some properties that its component parts lack, but also in the sense that systems continually give rise to qualitative novelty: new species, new ideas, new social systems. As Bunge writes: "In addition to developmental assembly processes, we must make room for evolutionary processes, that is, unique processes along which absolutely new things emerge." (2001A: 70). Systemism is thus a form of process philosophy, or as Bunge described it, emergent materialism.

Although Bunge's systemic approach emphasizes that events and processes were not entirely random, as if the world was some kind of cosmic casino, Bunge acknowledged that chance was an objective mode of becoming. Equally, although denying the validity of a rigidly deterministic world, Bunge emphasized that the world was indeed orderly and lawful, and that the principle of causality was therefore intrinsic to scientific understanding, as well as to everyday social life. As he put it, science is "deterministic in a lax sense" (2001B: 61). For Bunge, then, chance and necessity (causation), order (organization), and disorder (randomness), were all closely "intertwined" (2001A: 32).

If reality is radically "chaotic"—hyper or otherwise, as some philosophers seem to imply (Deleuze and Guattari 1988; Grosz 1988; Meillassoux 2008)—it is doubtful if life would ever have emerged on Earth.

Mario Bunge was an advocate of complexity theory, which implied that all material entities—whether physical things, organisms, people, artefacts, or social systems—were open, complex, dynamic systems; that all such systems have a history; and that all systems are dynamically linked in diverse and complex ways to other systems. The world we experience is lawful and orderly, but chance, randomness, spontaneity, and novelty were all also intrinsic aspects of the material world, from this point of view.

Although systems theory and the concept of emergence have been around since the 1920s or even earlier, many anthropologists have suddenly discovered the novelty of "chaos theory". It is however rather ironic that anthropologists who only a generation ago were extolling the virtues of literary theory, postmodernism, hermeneutics, and cultural relativism, denigrating and disparaging the Enlightenment, universalism, truth, and the empirical sciences in the process, should now be fervently embracing the highly deterministic, mathematical science of "chaos theory" (Mosko and Damon 2005).

But the idea that an earlier generation of anthropologists and social scientists—whether historical sociologists like Weber, Marxists, structural functionalists, world-system theorists, or cultural materialists—embraced the "totalizing determinism" of Newtonian mechanistic science and treated societies as closed physical systems (Mosko and Damon 2005: 8-9) is, however, just another postmodernist "caricature" of an earlier generation of social scientists and anthropologists. None of the above lacked a sense of history! (See I.M. Lewis 1999: 115-144; H.S. Lewis 2014.)

Two points need to be made here.

The first is that, as I described in chapter three, the classic Newtonian mechanistic paradigm began to be undermined during the eighteenth century with the emergence of the historical sciences—specifically geology, palaeontology, evolutionary biology, and the social sciences (including anthropology). Developments within the sciences during the nineteenth century, especially in relation to field physics and thermodynamics, as well as developments in chemistry and biology, further undermined Newtonian mechanistic science. By 1900, as Bunge noted, mechanistic science as a rigidly deterministic theory was "quite dead" (D'Albro 1939; Bunge 2001A: 37). The notion, then, that Newtonian mechanistic science only came to be critiqued around thirty years ago by systems theory (Capra 1982) and chaos theory (Gleick 1988) is therefore quite fallacious.

Secondly, the term "chaos theory" is, in many ways, a complete misnomer. For the scientific notion of "chaos" has nothing whatsoever to do with the idea of chaos as theorized by Plato, nor what we mean by chaos in our everyday life. That kind of chaos means disorder, utter confusion, without form, lawless, randomness, or unpredictability. But chaos in "chaos theory" simply means a critical dependence upon "initial conditions", and is, as John Gribbin writes, "completely orderly and deterministic, with one step following another in an unbroken chain of cause and effect which is completely predictable at every stage" (2004: 70).

To describe the world, both natural and human, as if it were a cosmic casino, to employ Bunge's apt phrase (2001A: 32)—that is, a "realm of instability, randomness and disorder" (Mosko and Damon 2005:8)—seems to me to completely misdescribe chaos theory. Implying also that postmodernism, existentialism, cultural relativism (historicism), and literary deconstructionism are *akin* to chaos theory—which is a deterministic science—because they are reflexive and "anti-deterministic", seems to me to be highly problematic—rather strained to say the least.

Bunge's complexity theory makes much more sense than Mosko and Damon's theory of chaos (disorder!), their take on a highly deterministic, mathematical, and reductionist science—"chaos theory".

In Defence of Systemism

Whatever is seen, Mario Bunge writes, has been looked upon from some viewpoint or perspective. There is, therefore, no "view from nowhere" (1999B: 4).

Systemism or systems theory, as envisaged by Bunge, was a synthetic (not holistic) and integral philosophy, a form of emergent materialism. It was a worldview, an "approach" to the world of nature in all its aspects, whether in respect to physical or chemical systems or with regard to such "emergents" as life (biology), mind (psychology), or social life and culture (anthropology and the social sciences). Advocating an *integral* perspective, Bunge was therefore critical of several alternative ontologies which he felt tended to have a "sectoral" or more limited approach to the understanding of the natural world and human social life. He specifically highlighted and critiqued five limiting perspectives, namely holism, atomism (individualism), structuralism, positivism, and environmentalism. I shall briefly discuss each of these in turn.

Holism

Holism, succinctly defined, accords primacy to the whole (from the Greek *holos*) of any system. It is the thesis that "the whole determines its parts, and that knowledge of the latter is unnecessary to understanding the totality" (Bunge 1999A: 122).

Holism is an ancient philosophy, reflected in all forms of animism (spiritualism). It was well expressed by Plato's idealist metaphysics, discussed in chapter two. In *Timaeus*, for example, the Greek philosopher described the world as a "living being with soul and intelligence" (1965: 42). During the late medieval period, holism came into its own among the advocates of natural magic, with their concept of a "world spirit" (*anima mundi*). It was given its modern expression in Jan Smuts' classic study *Holism and Evolution* (1926), for Smuts argued that there is an inherent holistic—"whole-making"—tendency within the universe, whether with respect to matter, life, mind, or the human personality.

But Bunge considered holism, as an ontological thesis, to be inherent in a wide range of *sociological* thinkers going back to Ibn Khaldun, all of whom tended to conceive of society as a "whole transcending its members". Bunge identifies this tendency, for example, in the conservative philosophies of Edmund Burke and George Hegel, the sociological theories of Comte and Durkheim, and the latter-day structural-functionalism of Radcliffe-Brown and Talcott Parsons (1996: 259). I shall return to the issue of sociological holism later in the manifesto.

Bunge (along with Martin Mahner) also considered the development systems approach in biology—that is, the notion of an "organism-environment system"—to be a form of holism. For besides being rather vague, it tends to underplay the importance and agency of the organism itself as a developing entity (Oyama 1985; Griffiths and Gray 1994; Mahner and Bunge 1997: 299-301).

As an epistemological thesis, holism puts a fundamental emphasis on the whole, specifically the social whole. It advocates theoretical intuition (*theoria*), implying that we can grasp the "whole" through spontaneous insight rather than through painstaking conceptual and empirical analysis. It is of interest that Bunge always linked holism, as a worldview, with nationalism, cultural relativism, the Gaia hypothesis, spiritual ecology, the idealist philosophies of Bergson and Husserl, new-age feminism, and, at extremes, with totalitarian political ideologies (Bunge 1996: 262, 2003: 103).

Bunge contends that the common expression "a whole is more than a sum of its parts", which is usually identified with holism, simply meant that a whole has properties that its parts lack. It is thus more an expression of Systemism, or emergent materialism, rather than of holism (Bunge 2001A: 30).

A holistic perspective seems to have been well expressed in recent years in the deep ecology of Fritjof Capra, and in Gregory Bateson's ecology of mind. Capra spiritualized ecology, equating systems theory with Asian religious cosmologies—the "teachings of the mystics"—given that they stressed the universal interconnectedness and interdependence of all phenomena, and the intrinsic dynamic nature of all reality (Capra 1982: 330). This is something, of course, recognized by all humans in their practical and common-sense understandings of the world! Capra seems to forget that religious mystics like Plato and Sankara denied the reality of the material world, reality being conceived as a spiritual realm or as pure consciousness (for a critique of Capra's spiritual ecology, see Bookchin 1995B: 215-216).

On the other hand, Bateson's (1972) holistic ecology views the mind as a "sacred unity" that embraces all living things. The mind consists of ideas or information—the "difference that makes the difference", as Bateson continually affirms—that are possessed by all living things, from redwood trees and the amoeba, in search of nutrients to human symbolic communication. Although clearly attempting to develop a systemic ontology that avoids the Scylla of religious mysticism and the Charybdis of Cartesian mechanistic philosophy—which he falsely equates with materialism—Bateson tends to view the "mind" as some disembodied entity that is "immanent" in all living beings (Bateson 1979: 28-29, 110; Bateson and Bateson 1987: 64).

The living world, for Bateson, is "joined together in its mental aspects"—the "sacred unity"—and he describes biological evolution not as a material or organic process but rather as the "evolution of mind" (1979: 29). He continually berates Darwin for ignoring the mind. But, of course, Darwin explicitly acknowledged that the senses, intuition, and various emotions, as well as such faculties as love, memory, curiosity, attention, imitation, and reason were found "in an incipient, or even sometimes in a well-developed condition, in the lower animals" (Darwin 1909: 193).

Bateson continually re-iterates Alfred Korzybski's (1949) well-known adage "the map is not the territory". Apart from some idealist philosophers and postmodern anthropologists, nobody, as far as I am aware, has ever confused a real bushpig foraging for roots with the *idea* of a bushpig!

Murray Bookchin has critiqued Bateson's systems view for its holistic and idealistic tendencies. He claimed that for Bateson "materiality is dissolved into inter-relationships and then subjectivized as minds", thereby divesting nature of its "very physicality" (Bookchin 1995A: 114; on Bateson's ecology of mind see also Harries-Jones 1995; Charlton 2008).

Atomism (Individualism)

Atomism, the notion that the world consists of material atoms moving in a void, was advocated by ancient Greek and Indian materialists (discussed above). It is perhaps the earliest naturalistic worldview. This theory, though initially qualitative and speculative, was later confirmed experimentally by later physicists such as Dalton and Einstein (Bunge 1999A: 25, 2001A: 34).

Atomism is essentially the ontology underlying individualism, the thesis that every whole is nothing but an *aggregate* of its parts. Individualism is thus an exact antithesis of radical holism, and was embraced by numerous scholars from diverse intellectual backgrounds, including, Bunge suggests, Thomas Hobbes, Jeremy Bentham, Max Weber, William James, and Alfred North Whitehead. All those scholars affirm that "all existents are individuals, whereas all wholes are conceptual" (2003: 84).

Ontological individualism (or atomism) gives primacy to the individual, denying the existence of emergent properties. In terms of social theory, it denies the existence of social bonds and social systems, or asserts that these are fully reducible to individuals and their actions (1996: 243). The epistemological counterpart to atomism is methodological individualism, which stresses that primacy be accorded to the individual in the human sciences, and holds that an understanding of all social facts and all social organizations can be achieved solely through the explanation of the observable actions of individuals (Bunge 1996: 245).

Individualism, Bunge contended, was a popular approach in philosophy and social studies precisely because it was a *reaction* against the excesses of holism. Bunge offered many telling critiques of the (micro-) reductionism and one-sidedness of many forms of individualism, exemplified by the following intellectual projects: physicalism or eliminative (reductive) materialism (Churchland 1984); sociobiology and evolutionary psychology (Wilson 1975; Dawkins 1976; Cosmides and Tooby 1987); the rational egoism of Ayn Rand (1964); social behaviourism (Homans 1974); and, finally, rational choice theory and neo-classical economics (Becker 1976) (Bunge 2003: 149-163).

The problem with individualism is that it fails to recognize that molecules, organisms, and social systems, for example, have an integral reality; they are a *structured* collection of parts, not a mere aggregate, and that they cannot therefore simply be reduced to their parts. Bunge emphasizes that (micro-) reductionism and individualism are methodological "partners", and indeed affirms that every entity, apart from the universe as a whole, is a component part of some larger system. Individualism is therefore fundamentally at odds with systemic theory, for it is opposed to the concepts of both emergence and levels of organization. It is a flat ontology (Bunge 2003: 140).

Structuralism

In contrast to holism, which focusses on the totality, and individualism, which stresses the ontological salience of the individual (part), structuralism emphasizes the structure, or internal relations, of a material entity or system. The problem is that the structure of a system tends to be emphasized at the expense of its composition and its environment. But Bunge always affirmed that every structure (or relation) is the structure of *some*thing, and that there are no structures in themselves, no free-floating or autonomous relations. A structure then is the property of some system, and is a set of relations among its component parts, especially those which hold the system together as an integral entity (Bunge 1999A: 277, 2001A: 42).

Bunge identified structuralism not only with Karl Marx, who famously described the human individual as an "ensemble of social relations" (Marx and Engels 1968: 29), but also with the writings of Claude Levi-Strauss (1963) and Pierre Bourdieu (1968). To define the individual—whether a material thing or a human person—as a mere "set of relations", is, Bunge argued, quite misleading, for there are no relations without *relata* (things). As we indicated in chapter five, many scholars (for example, Whitehead, Bateson, and Ingold) have a tendency to define—reduce—material things to a "knot" or "collection" of relations. But for Bunge, our experiences of the world are less experiences of abstract "relations" than of concrete material things in constant interaction and change. Colour, for example, is the property of *both* a material object in the world, and of an organism whose perceptions and nervous system analyse the image thrown upon its retina: colour involves more than just a "relation". No wonder Bunge described the structuralists—the relational epistemologists—as committing a "logical fallacy" in defining individuals (or groups) merely as a set of relations. As he put it, to assert the ontological primacy of *either* a relation or its relata (material thing) is to "misconstrue the very notion of a relation" (1996: 273, 2003: 82).

As Bunge concluded:

Both in logic and in science, individuals and properties—whether intrinsic or relational—come together on the same footing; neither is prior to the other. In particular, there are no relations without relata—by definition of "relation". Moreover, every entity emerges and develops in interaction with other entities. This holds for persons and corporations, as well as for molecules, cells and other concrete entities. (2003: 86)

Positivism

I have, in the last chapter, discussed Bunge's critique of positivism, in particular its phenomenalist metaphysic and its empiricist theory of knowledge. Positivism, of course, is closely associated with both individualism and with (micro-) reductionism, a research strategy whereby a complex system is explained by reduction to its constituent parts.

What specifically troubled Bunge with regard to positivism is that it focussed exclusively on the observable world of actual experience, the world as it appears to humans. It is therefore a form of descriptivism, denying that there are any underlying mechanisms that may explain phenomena.

Positivism thus ignores the importance of "mechanisms", the underlying relations and processes within some material entity that give rise either to its emergence, its functioning, or—to employ Bunge's own favourite expression—what makes a thing "tick". Thus, a mechanism is "whatever process makes a complex thing work". The disclosure of such mechanisms goes beyond description or inductive generalizations (positivism), in providing an explanation of phenomena, whether natural or social (Systemism) (Bunge 1999A: 173, 2001A: 41).

Environmentalism

Given that all material things exist in a "world of systems", all things are situated in a "web" of relationships. They can, therefore, only be fully understood if placed in a wider context—its environment. Just as holism (macro-reductionism) and individualism (micro-reductionism) each have a certain validity if not taken to an extreme, reference to the environment of a thing—its context—is not only important in understanding the phenomena, but unavoidable (Bunge 2003: 130).

Problems arise in over-emphasizing a system's external environment—*radical environmentalism*—thus overlooking its composition, structure (internal relations), and mechanisms, as well as the integrity and agency of the system itself (Bunge 2003: 39). For Bunge, such environmentalism was particularly evident in many sociological studies of science, psychological behaviourism, various forms of environmental determinism, and those Marxist theories that tend to downplay human agency (2003: 131).

In a series of substantive and critical studies, Mario Bunge applies his systemic approach to a wide range of phenomena and issues. In relation to the present manifesto, of particular interest are Bunge's systemic writings on organic life, on the mind/body distinction, and on human social life and culture. Bunge invariably takes a triadic—dialectical—approach, one that mediates between the epistemological extremes of radical individualism (positivism) and holism (idealism) to elicit and explore a more viable systemic approach—emergent materialism.

With respect to a systemic approach to *organic life* (biology) Bunge attempted to avoid two extremes: mechanism and vitalism.

Mechanism in biology is a form of reductive materialism (physicalism) that tends to define and interpret life in terms of physico–chemical elements, an approach well expressed in the "gene-centred" perspective of the Neo-Darwinists. But Bunge emphasized that biology has demonstrated that organic life has a number of emergent properties and processes that are simply unknown to physics and chemistry (Bunge 2001A: 56; Mayr 2004).

Vitalism, on the other hand, which Bunge considered a descendent of animism (religious thought), treats life in terms of some immaterial entity such as *entelechy* (Hans Driesch), or some *elan vital* (Henri Bergson). He considered vitalism to be a variety of spiritual holism, and such concepts as *entelechy* to be imaginary and inscrutable, and thus "beyond the ken of science" (2003: 46). The evolutionary biologist Ernst Mayr likewise considered that vitalism, in all its forms, had been "totally refuted" by contemporary biology (1982: 131).

Yet, in recent years, vitalism has been given a new lease of life, mainly through the writings of Gilles Deleuze as well as through the best-selling books of the vitalists Nietzsche and Bergson. Deleuze has, in fact, been described as a vitalist materialist.

I have discussed Deleuze more fully elsewhere (2014A: 693-723) but two points may be made in the present context, specifically with regard to his vitalist ontology.

The *first* is that Deleuze (and Guattari) tend to give the concept of "life" the most obscure and abstruse meanings. Sometimes they appear to conflate "a life" with singular events—a day, a season, a wind, a fog, a swarm of locusts, even an imaginary vampire out at night (whither Buffy the Vampire Slayer?!) (Deleuze and Guattari 1988: 262). At other times, life is identified with the so-called "body without organs"—unorganized matter—and with the smooth, molecular side of the "primordial duality" that pervades their work. They thus write that "every thing is alive", describing life as "inorganic, germinal and intensive" (1988: 499). This leads them to describe organisms or living beings—whether cabbages, grasshoppers, bushpigs, or humans—as striated, molecular entities that are "opposed", or even the "enemy" of life—equated with the inorganic or physical "body without organs". Indeed, organisms are perversely viewed not as a manifestation or expression of life, but as curtailing or inhibiting the "flow" of inorganic life (Deleuze and Guattari 1988: 158). Deleuze and Guattari seem to confuse life with energy. If everything is alive, how do we account for the demise of organisms?

Secondly, this vitalism or life philosophy of Deleuze (and Guattari) is expressed, ironically, in the most mechanistic idiom. For they stridently affirm the universality of mechanic assemblages and connections. They describe the world as a mechanosphere. But, of course, they deny that their account is a representation or image of the world (what then is it?), preferring to identify themselves with the plane of consistency, the body without organs, the rhizomatic sphere, the nomads on the Eurasian steppes, the virtual realm, chaosophy, or informed matter (take your pick!), or, more simply, to embrace the flow of *in*organic life (Deleuze and Guattari 1988: 1-74). Such a vitalist ontology seems to me to be a form of reductive materialism in which organic life, in its varied forms, and human social existence are reduced to machinic assemblages of unorganized matter.

What Deleuze and Guattari had in common with Mario Bunge was an almost complete rejection of hermeneutics and phenomenology, or any emphasis on lived experience.

But although Deleuze and Guattari were anti-state, they denigrated, like Nietzsche, history, reason, and truth, while extolling the virtues of de-territorialization, nomadism, chaos, and fragmentation (schizophrenia). It is therefore small wonder that scholars have described Deleuze as a postmodernist, and as the "ideologist of late capitalism" (Zizek 2004: 184; Bunge 2012: 8). The neo-conservative Roger Scruton is even more dismissive, describing Deleuze as an intellectual "fraud", and his philosophy, like that of Althusser, Lacan, and Derrida, as an effusion of academic "nonsense" (Scruton 2015: 180-196; but c.f. my own critical engagement with Deleuze's vitalist materialism, Morris 2014A: 693-723).

Deleuze also continually disparaged the integral unity and self-organization of organic beings, including the human subject—thereby downplaying human agency—and although he (and Guattari) recognized the existence of several ontological "strata"—the physico-chemical, organic, and social (alloplastic) levels (1988: 502)—unlike Bunge and the dialectical biologists discussed in the last chapter, Deleuze was not an evolutionary (historical) thinker. Deleuze instead emphasized only the "becoming" of things, including imaginary vampires (1988: 238). What Deleuze refers to as the virtual realm and the morphogenetic processes of actualization (1994: 211-215), Bunge and Bhaskar describe more simply as "mechanisms" within material things—dynamic systems (for further critical insights into Deleuze's ontology, see Badiou 2000; Hallward 2006).

Mario Bunge's systemic, materialist approach, in contrast to both mechanism and vitalism, considers life to be the property of material things of a special kind—living beings—and purposeful behaviour as exclusive to birds and mammals (2001A: 56).

It is noteworthy that Bunge, unlike many contemporary biologists and philosophers, held that a species is essentially a biopopulation, not an individual or concrete system extending over time and space (Bunge 2003: 48; Mahner and Bunge 1997: 154; c.f. Delanda 2006: 48-49). Bunge thus affirmed that the organism in its environment is the basic unit of life, and was sceptical of the concept of "autopoiesis". He regarded this term as just a fancy synonym for both "self-organization" and "self-maintenance" (Mahner and Bunge 1997: 144).

In his many writings on the perennial *mind/body problem* and on the nature and scope of psychology, Bunge was critical of two contrasting tendencies. These he graphically described as mindless neuroscience and brainless psychology, although cognizant of the complexity of the issues involved.

The first approach tends to deny the reality of mental life. It does so either, as with behaviourist psychology, by dispensing with the concept of mind altogether—explaining human behaviour in terms of external stimuli—or through various forms of reductive materialism. The latter approach, exemplified in extreme by eliminative materialism (Churchland 1984; Stich 1999) tends to reject such phenomena as "thought", "emotions", "memory", "imagination", and "feelings", regarding them simply as "folk" psychology, and identifies all mental activity with neurophysical processes. Both approaches are, of course, at variance with the fact that the psychological sciences over the years have produced many insightful and illuminating studies of such phenomena as cognition, learning, memory, the emotions, and human perception. Eliminative or reductive materialism—*mindless neuroscience*—along with behaviourism, are thus both untenable. Both approaches are forms of positivistic science that essentially involve *reducing* psychology to biology.

The second approach—*brainless psychology*—is particularly associated with Cartesian philosophy and entails a psycho-physical dualism. This is the oldest of all philosophies of mind, and derives essentially from early shamanic thought and most forms of religion. This conception of the mind—which views it as immaterial and completely separate from the body—was incorporated into philosophy, Bunge, contends, by Plato and Descartes.

The doctrine of the immateriality of the mind, though long critiqued, continues to be expressed, according to Bunge, in many contemporary modes of thought. He particularly highlights, for example, psychoanalysis—Freud's concepts of id, ego, and superego, which he adjudges to be rather ghostly entities—the cultural idealism of many interpretive anthropologists, who treat symbolic forms as completely autonomous from social life; peoples' material relations with the world; and a great deal of contemporary cognitive science. Many cognitive psychologists associated with the computational theory of the mind (e.g., Dennett 1991; Pinker 1997) tend to treat mental processes as completely unrelated to biological phenomena. Such mental processes consist of information processing that may be "embodied" either in fleshy bodies or machines. Bunge describes such cognitive scientists and advocates of artificial intelligence as embracing a form of "mentalism", that ignores the relationship of the mind to the human nervous system (Bunge 2001B: 265-280).

In contrast to eliminative materialism and behaviourism (which dispense with the mind), and the functionalism of the cognitive scientists (which treat the mind as some immaterial entity completely separate from the body), Bunge advocates a "biopsychological" approach that emphasizes that the human mind has evolved alongside its organ—the brain. Always paying tribute to the seminal writings of Donald Hebb (1980), and drawing on the findings of contemporary research in neuroscience, Bunge advocates an approach that he describes as cognitive neuroscience. He highlights the following:

- That the mind is a set of mental states of processes—whether relating to perception, cognition, emotions, feelings, memory, imagination, or volition—that are essentially functions of the brain (the nervous system);

- The mind, along with the brain, has emerged during the course of human evolution, but, unlike Bateson, Bunge views the mind as essentially restricted to vertebrates, particularly birds and mammals;

- That the mind (minding) is what the brain does, and the relation of the mind to the brain is akin to that between walking and the legs;

- That the brain is an extremely complex organ, and not only consists of several sub-systems, but is characterized by an inherent plasticity; And, finally,

- Bunge emphasizes that the mind, as the property of an organism (or person) is always situated within a social context, and this context has to be taken into account in the understanding of human behaviour and subjective experience.

Psychology must therefore forge links with both biology (especially neuroscience and evolutionary biology) and the social sciences. But Bunge warns against over-stressing the social context, as with the behaviourist psychology of B.F. Skinner (1953), the social behaviourism of Lev Vygotsky (1978), and the ecological psychology of James Gibson (1979)—all of which, according to Bunge, tend to ignore the human brain (Bunge 2001A: 96, 2003: 49-52, 2010:143-199).

In several seminal texts linking philosophy with the social sciences, Bunge (1996, 1998) also applied this systemic approach to the understanding of human social life. I shall draw upon some of Bunge's ideas and insights on social systems later in the manifesto.

A scholar of wide erudition, Mario Bunge attempted to construct an *integral* philosophy, one that combined the following: realism, emergent materialism (an ontology akin to what Murray Bookchin described as dialectical naturalism), a ratio-empiricist epistemology, Enlightenment humanism (which I described in chapter three), and a deep attachment to empirical research and to the scientific method as a way of understanding both the natural world and human social life and culture (Bunge 2012: 15-34).

By humanism, of course, Bunge did not mean Cartesian philosophy and an ethic implying the human domination of nature—as quite falsely defined by postmodern literary scholars like Braidotti (2013)—but rather the search for truth and justice. Humanism, Bunge wrote, entailed the condemnation of military aggression and racism, as well as all forms of political oppression and cultural exclusion; the upholding of such universal values as solidarity, equality, and freedom; and a materialist outlook that encourages the pursuit of knowledge through sense experience, reason, and the imagination (Bunge 2001A: 14-15, 2012: 26-27).

But, given his background in mathematics and theoretical physics, Bunge never seems to have envisaged Systemism as implying an ecological worldview and sensibility. Further, given his vibrant and staunch emphasis on the empirical sciences—and scientific realism—Bunge expressed very little sympathy for phenomenology and hermeneutics, which are intrinsic to any understanding of human societies and their symbolic cultures. Both constitute limitations with respect to Bunge's systemic philosophy.

Nevertheless, Bunge's Systemism (emergent materialism—a form of complexity theory), along with the evolutionary (historical) materialism that stems from Darwin and Marx, seems to me to provide the essential foundations for the advocacy of anthropology as a historical science. It is to this task I shall now turn.

PART THREE

ANTHROPOLOGY AS A HISTORICAL SCIENCE

Chapter Ten

Truth and Representation

As I discussed in the first chapter, one of the main tenets of postmodern scholars such as Heidegger and Rorty, along with postmodern anthropologists, was a repudiation of both truth (as correspondence to reality) and the concept of representation. Any ethnography that aims to provide a realistic and illuminating account of the social life and culture of "the other" (as postmodernists perversely describe people of other societies), rather than engage in poetic revelry and fashionable autobiography, and any conception of anthropology as an empirical historical science, of course pre-suppose not only a realist ontology but the concepts of truth and representation. This chapter is therefore devoted to explicating and defending the integrity and salience of these two essential anthropological concepts.

The Correspondence Theory of Truth

Ontological realism, the notion that the material world, or nature, *exists* independently of human cognition is practised, Bunge suggests, by "all sane people" (2006: 250).

But realism is the *bete noire* of all postmodern theorists, whether they take their inspiration from phenomenology, philosophical hermeneutics, or Neo-Kantian cultural idealism (aka social constructionism). In embracing a phenomenalist metaphysic, logical positivism also rejects critical realism—as do all radical empiricists, as discussed in chapter eight.

It is therefore hardly surprising that postmodernist scholars, whatever their complexion—whether anthropologists or literary theorists, whether historians or devotees of Heideggerian philosophy—have tended, in nihilistic fashion, to reject not only the concepts of nature or thing (invariably placed in inverted commas as if they had no reality), but also such concepts as reason, history, society, and class. Some postmodernists even proclaim the "end", or the "dissolution", of philosophy and anthropology. But there are two concepts in particular which tend to be denigrated, or even rejected completely, by postmodernists and Neo-Kantian scholars more generally: truth (as correspondence with a material reality) and representation.

Two points of clarification, therefore, need to be made at the outset. Firstly, it is important to recognize that there are many different kinds of truth. There are the truths of mathematics and formal logic, and there are artistic truths, poetic truths, semantic truths, moral truths, and the truths of religion (Bunge 2006: 193-194). That Jesus—given that he was the son of God, according to Christian doctrine, and thus had spiritual powers beyond those of ordinary mortals—was able to walk on water and feed a multitude of people on but two loaves of bread and five fish, is a religious truth. Such religious truths may be manifested or expressed in many ways: through gnosis or mystical states, through divine revelations as recorded in holy scriptures, or through the oracular pronouncements of spirit mediums or religious gurus.

Finally, of course, there are the factual or empirical truths of science and everyday life, and of practical knowledge. These are truths about the material world and human social life that involve what is usually described by philosophers as the correspondence theory of truth. This conception of truth, adopted in our common-sense understandings of the world and by the various empirical sciences, presupposes a realist ontology.

The idea that "man invents his own realities" (Wagner 1981: ix) is an idealist fantasy, or at least a crude exaggeration. If that were indeed true, we would no longer envisage erupting volcanoes destroying people and property! The various interpretations and explanations of such (real) events are, of course, a human invention—it was due to the wrath of God—but not the erupting volcano! That symbolic culture is also a human creation or "invention" is clearly not some profound postmodern insight: it is a truism! Roy Wagner is therefore hardly a "genius" for telling us that culture is a human invention, whether with respect to the culture of the anthropologist or that of the people they study (Holbraad 2012: 270). No anthropologist, as far as I am aware, claimed to be divinely inspired, nor that cultural ideas were not created by humans and instead emanated from some deity or the ancestral spirits—whatever people themselves may hold in terms of their own religious (animistic) cosmology.

Please note an important contrast: while scientists, including realist anthropologists like myself, attempt to discover truths about the world with respect to both the natural world and to human social life and culture—truths that are always provisional, and open to critique and revision—religious devotees instead proclaim *the* truth, as something that is not to be questioned. Given that such truths are derived from a spiritual or transcendent source, they constitute what the ecological anthropologist Roy Rappaport described as "absolute truths" or "eternal verities" (1979: 119; Morris 2006: 245-246).

One hardly needs to study esoteric religious cults to realize that, in *all* human societies, empirical truths recognized by everyone co-exist, and are different from the truths expressed in religious myths, doctrines, and ritual practices.

Secondly, it is important to recognize that the concept of "representation" has two very contrasting meanings. (Although, of course, idealist philosophers and their various postmodern acolytes have, by definition, no use of the concept of representation, as they confuse or conflate ideas—the map—with the material world—the territory!)

On the one hand, the concept of representation is identified with the classical empiricism of John Locke and the kind of empirical (or naïve) realism I described in chapter eight. This is how both Gilles Deleuze (1994) and Richard Rorty (1980) interpret—and so dismiss—the concept of representation. For it implies that a representation is a "mirror of nature", that it is simply a "reflection" or "image" of the material world. It entails what Bunge refers to as the "mirror theory" of knowledge (2010: 186). As we noted in chapter eight, Wittgenstein's "picture theory" of language and Lenin's version of empirical realism both express this conception of representation as a faithful *copy* of a real external world. This conception of representation has long been critiqued by materialist philosophers, psychologists, and realist anthropologists.

On the other hand, the concept of representation is used by philosophical materialists and many realist anthropologists (including myself) to refer not only to films, books, and maps, but also the following cultural phenomena: collective representations (Durkheim), political ideologies (Marx), philosophical worldviews (Bunge), scientific theories (Hacking), cultural configurations (Benedict), religious cosmologies (Holbraad—no less!), images of thought (Deleuze), and symbolic forms (Cassirer) (Morris 2014A). Such representations, like all forms of knowledge—as conceptual systems—are always *about* something—aspects of the material world (including human social life and culture). They do not, of course, have an autonomous or independent existence (nor are they unchanging), but always presuppose, and are situated within, specific social-historical formations. The relationship between a representation and the material world—whether natural or social—is not pictorial but symbolic, and, if factual or scientific, may be expressed by such concepts as mapping (Bateson 1979) or cartography (Deleuze and Guattari 1988).

Thus representations "are not a copy of the world but a symbolic construction of it", and they always imply a triadic ontology involving the identity of the human person (the knower), the social context, the diverse systems of social relations, and a "mode of relating to the world outside" (Jovchelovitch 207: 20).

The film "Star Wars" is, of course, a representation—of a world of the human imagination. But people are fundamentally earthly creatures. They live in a real world, not just in the world of the religious or artistic imagination, nor in any of the "possible worlds" conjured up by detached esoteric philosophers (D. Lewis 1986; see Bunge 2010: 162 for a critique of such "fantasies").

It is therefore quite misleading to equate—as many Neo-Kantian idealist scholars do—representations, whether of a real or imaginary world, with the correspondence theory of truth. The latter is a semantic category relating specifically to *factual* or empirical truths. All representations, it hardly needs saying, are socially mediated, and have an impact upon the lives and actions of human beings in all societies. Deep ecologists and Gregory Bateson (1979), for example, relate the present ecological crisis to the fact that people still embrace and advocate the Cartesian mechanistic philosophy of the seventeenth century.

It is of interest to note that, although fervently claiming to reject the concept of representation, Richard Rorty (1980) offers us a representation—an interpretation—of the philosophies of Dewey, Wittgenstein, and Heidegger; Deleuze and Guattari in *A Thousand Plateaus* (1988) present an esoteric cosmic vision and representation of the world as a chaotic "mechanosphere"; and, finally, Martin Holbraad's *Truth in Motion* (2012) offers us a representation of Ifà divination in Cuba and its religious practitioners—although these scholars ironically claim to renounce the concept of "representation"! Truth in motion? Truth, of course, is a concept; it is not a material thing, it has no energy, and it certainly does not move around. Only people move around and engage in disputes over issues of empirical validity and truth.

Both the religious and correspondence theory of truth, of course, allow for many things and events in the world to be described as true.

Although offering us an interesting account of Ifà divination in Cuba, Martin Holbraad—with some pretension—advocates a completely new approach to anthropology, dubbed "recursive anthropology", arrogantly dismissing current anthropology as an intellectual "mess". The term "recursive" simply indicates what scientists—including biologists, psychologists, anthropologists, and other social scientists—have been doing for a century or more: creatively developing theories (representations) and hypotheses that accord with the material world—both natural and social—in which humans find themselves. But social scientists and realist anthropologists are not just concerned with providing an "adequate description" (representation) of social reality (Strathern 1988: 10-12), or with interpretative "analysis". They are instead most interested in explaining social and cultural phenomena by reference to underlying mechanisms and processes, especially situating cultural phenomena, like that of religion, in its wider socio-historical context (Morris 2006).

Extolling the postmodern concept of "alterity", Holbraad suggests that whatever people may do, say, or think, we—that is, anthropologists—are not "equipped to represent it" (2012: 246). But the religious ideas of people in Cuba or in remote parts of Amazonia are no more exotic and beyond understanding than Tibetan Buddhism, evangelical Christianity (and other forms of Christian animism), the varied form of Islam or Hinduism, African religious cosmologies, or Nazi ideology. Needless to say, not only anthropologists but people in all societies offer representations—descriptions, interpretations, and explanations—not only of things and events in the natural world, but of people, social activities, and cultural phenomena—both of their own and those of other cultures. For almost a hundred years, anthropologists have been doing what Holbraad suggests can't be done, namely providing us with insightful and often illuminating accounts—representations—of other societies and their life-ways. Moreover, describing the history of anthropology only in terms of three paradigms—evolutionism, diffusionism, and social constructionism—as Holbraad does (2012: 19-34)—is quite facile given the rich tapestry of anthropological history and theory (on the diversity of theoretical approaches to religion, for example, see Morris 1987; on the complex history of anthropology see Stocking 1992, 1996; Eriksen and Nielsen 2001).

What comes strongly through Holbraad's theoretical musings is a marked aversion to the correspondence theory of truth (while using it!) and an anti-realist perspective—Holbraad simply following in the wake of Heidegger, Bruno Latour, and the postmodernists.

Misleadingly equating the concept of representation (as a system of meanings) with the correspondence theory of truth, denying that we can in fact "represent" other people's social life and culture, and seemingly rejecting the concept of representation, Holbraad—lo and behold!—admits in the conclusions of his study that his own account of Ifà divination in Cuba is in fact a "representation" (no less!). He even alludes to its "accuracy" as a representation (Holbraad 2012: 250), thus accepting the correspondence theory of truth that he continually denigrates and explicitly rejects!

Going beyond representations—descriptivism—is what, of course, social scientists and realist anthropologists have been doing for many decades.

Like many postmodernist scholars, who are inordinately fond of inventing neologisms—as brand names?!—to describe sociological commonplaces, Holbraad describes his own approach as "ontographic". He thus gives the false impression that his own approach is realist and entails a materialist (naturalistic) ontology. In fact, his approach to anthropology is essentially Neo-Kantian, a heady mixture of combining positivism (phenomenalism), social constructionism, and mysticism.

It is indeed rather ironic that while Gilles Deleuze and many postmodernists (including Holbraad) are adamant in rejecting the concept of "representation", other scholars are telling us that all animals respond to "signs" in their environment, that all life-forms "represent" the world in some way or other, and that "representations" are intrinsic to their being (Kohn 2013: 9).

The classical theory of truth as correspondence has been taken for granted by ordinary people of all cultures, and by most philosophers from Aristotle to Bunge. All theories are representations in the sense that they represent reality, whether natural, social, or as it is "in-itself" (Harre 1970: 14; Manicas 2006: 3). Knowledge thus consists of a search for truth—a vocation now derided by some philosophers (like Deleuze) and by most postmodern anthropologists. It is a form of cognition shared by the empirical sciences and everyday practical knowledge. As Hannah Arendt suggests, it is quite distinct from "thinking" (philosophy), which goes beyond what is known, and is fundamentally concerned with meaning and with how human life should be lived (ethics). Nothing but confusion reigns if science and philosophy are conflated, and, as with positivism and Heidegger, a "basic fallacy" is committed by interpreting "meaning on the model of truth" (as correspondence) (Arendt 1978: 14-19).

The classical definition of truth was expressed by medieval scholars as *Veritas est adequate rei et intellectus*—"truth is the agreement of knowledge with its object (things)". That this is what *constitutes* truth—not, as with coherence or pragmatic theories, what is the *criteria* of truth—has been acknowledged by philosophers throughout the ages: Aristotle, Ibn Sina, Aquinas, Kant, Husserl, Popper, Davidson, and Bunge (c.f. Husserl 1970: 176; Davidson 1984: 37; Popper 1992: 5). Even pragmatist philosophers such as Charles Peirce and John Dewey (contra Rorty) affirmed the correspondence theory of truth. Both were realist philosophers, and Dewey, who was deeply influenced by Darwin, suggested, for example, that a true idea in any situation "consists of its agreement with reality". He emphasized that a sense of correspondence was operational, and that the relationship between an idea and an actual state of affairs could only be verified through enquiry and practical action, such as observations (Dewey 1916: 150-151). Likewise, Peirce affirmed that "Truth consists in the existence of a real fact corresponding to the true proposition." (Buchler 1955: 160; on American pragmatism see Thayer 1981; West 1989; Morris 2014A: 140-183).

Replacing truth with the notion of "consensus", as does Rorty (1991) is quite facile. Although Rorty continually described Dewey as his "hero", and often in his own writings employs the phrase "we pragmatists", Rorty not only abandons but also explicitly repudiates the main tenets of Dewey's philosophy: his Darwinian empirical naturalism, his affirmation of ontological realism and the correspondence theory of truth, his stress on the cogency of scientific enquiry, his ethical naturalism, and, finally, Dewey's radical politics! As Boisvert suggests, Rorty is not a pragmatist but simply an advocate of the then-fashionable postmodernism (1998: 176; Morris 2014A: 160).

Likewise, in embracing radical empiricism (positivism), and in repudiating the correspondence theory of truth and representation, Michael Jackson (1989: 14), following Rorty, completely misunderstands John Dewey's philosophy. For Dewey clearly recognized that pragmatic (instrumental) truths presupposed the correspondence theory of truth, strongly affirmed a *scientific* approach to social life, and described his own philosophy as a form of empirical *naturalism* (Dewey 1925).

The correspondence theory of truth affirms that a proposition, whether expressed as a statement, theory, or representation, is *factually* true if it "fits", "corresponds", or is "adequate" to the facts to which it refers. Such facts (things and events in a real world) can only be verified, as Bunge, like Dewey, affirmed, by recourse to *empirical* operations. This entails engaging with the material (or social) world by means of observations, empirical testing, experiments, or, with respect to anthropology, by participant observation. Humans do not have direct access to the world through some kind of mystical intuition—whatever mystics or gnostics may say—but grasp or understand it only through sensual experience and reason. Experience entails our interactions with the world, whether through perceptions, feelings, actions, or, with respect to social life, through interpretive understandings (*verstehen*).

Thus, for example, the statement "it is raining" is true only if it is in *fact* raining, and this can be established only by empirical observations or simply getting wet. The confirmation of a scientific theory is, of course, far more complex, as it often involves facts that are not directly observable. Yet this still entails *empirical* testing, or evidence, to establish the validity of the theory or hypothesis. What has to be recognized is that the correspondence theory of truth refers only to *factual* statements, whether with respect to everyday empirical knowledge or with respect to the theories of both the natural and social sciences, including anthropology (Bunge 1996: 167, 2006: 257-263).

The correspondence theory of truth does not entail suspending the testimony of the sense (Plato, Descartes); nor does it lead to "infinite regress" (Holbraad 2012: 206); still less does it entail bringing in the deity to give us a "god's eye view" as to whether the relationship between a statement (proposition) and the facts (events in the world) really does hold! Sensual experience and the human brain suffices! Doubting whether it is raining, or a volcano is really erupting (and is not just a cultural invention), indicates the degree to which Neo-Kantian cultural idealists have become alienated from nature—the material world. Holbraad suggests that his own account of truth may have an air of "vacuous mysticism" (2012: 205). This is indeed the case, for in rejecting both the concepts of truth (as correspondence) and representation, Holbraad is rather like the proverbial naïve lumberjack who is engaged in cutting off the branch on which he is sitting! For he continually provides us with a representation of Cuban divination, and leads us to presume that his account is indeed true as a representation of this cultural phenomena—not just a figment of his own imagination like Star Wars and Buffy the Vampire Slayer. As I noted earlier, Holbraad thus reluctantly admits to this in the conclusion to his study (2012: 250). Needless to say, the very idea of making a mistake (or error), or recognizing a fake, presupposes a realist metaphysic and the classical conception of truth (Trigg 1980: xx).

It is misleading, indeed obfuscating, to conflate, as many hermeneutic and postmodern scholars do, the correspondence theory of truth and the objective theory of meaning, which assumes an isomorphic or mimetic relationship between consciousness (or language) and the material world. It is equally obfuscating to equate the correspondence theory of truth (or realism) with "absolute truth" (Carrithers 1992: 153; Tallis 2012: 39-41). Falsely implying that ontological realism and correspondence theory entail "absolute truth" or "absolute foundations", many early postmodern anthropologists, as I discussed in chapter one, tended to "flip" to the other extreme. They thus came to embrace extreme cultural relativism and an anti-realist metaphysic that implied some form of Neo-Kantian cultural idealism, usually described as social constructionism. They thus reject the concept of truth as correspondence—thirty years later Holbraad follows in their wake—and deny any kind of relationship between ideas (or thought), whether or not expressed as representations, and the external material world. Julia Graham expressed this form of cultural idealism rather cogently: "Thought processes are *sui generis* and are completely determined rather than a representation of something else. They cannot be validated on the basis of correspondence to the '*real world*'. The criteria for validating some thoughts rather than others are internal to a theory." (1990: 59-60; Sayer 2000: 69).

Scientific materialism, and especially cognitive neuroscience (psychobiology), has demonstrated that this form of cultural idealism, which resurrects a Cartesian radical dichotomy between cultural configurations (representations) and the material world, is completely untenable.

Some critiques of the correspondence theory of truth seem quite bizarre. Misleadingly, correspondence theory is equated with Cartesian dualistic metaphysics, falsely implying that it involves no humans at all, Lakoff and Johnson conclude—in self-contradiction—that "A person takes a sentence as 'true' of a situation if what he or she understands the sentence as expressing *accords* [sic!] with what he or she understands the situation to be" (1999: 106).

Though oddly expressed, this is precisely what most people mean by the correspondence theory of truth, though why these scholars need to put "true" in inverted commas remains obscure.

Thus the radical dichotomy proposed by Martin Heidegger (1975), and avidly embraced by his postmodern acolytes (Richard Rorty, Marilyn Strathern) between thinking (evocation, lived experience)—acclaimed—and metaphysics (representation, scientific knowledge) is quite vacuous. Contrary to ultra-rationalism and theology, knowledge (representation) is always based on our engagement with the world (whether natural or social); while, contra radical empiricism, our responses to the material world (evocation) always entail prior knowledge about the world. Life and thought are intrinsically interconnected in a complex dialectic. Heidegger, of course, expressed his own metaphysic as a form of mystical theology.

Having introduced and defended the concept of truth (as correspondence to an ever-changing material reality—the facts) it may be useful to discuss the concept of meaning.

Truth and Meaning

The fact that the world—the Earth—exists and has powers to act independently of humans, that humans are relative late-comers in the history of life on earth, and that human life is but a "spec" in a vast expanding universe, has led some scholars to suggest, in nihilistic fashion, that the world for humans has no meaning (Brassier 2007). This is patently not the case—especially for emergent materialists. For the world is not only intrinsically meaningful to humans, but for all living organisms. Plants respond to "attacks" by herbivores, whether insects or mammals, by producing toxins to deter the intruders; bacteria and various microbes are able to feel, or are aware of, magnetic fields and the presence of both potential nutrients and harmful forces in their environment; and all animals, but especially birds and mammals, respond to what Jesper Hoffmeyer (1996) describes as the "signs of meaning in the universe"—responding, that is, to what Bateson (1979) refers to as "information". Such signs may relate to vision, chemical signals, thermal radiation, smells, or sounds. As Hoffmeyer puts it: humans and other organisms live in a world of signification—or meaning—and "Everything an organism senses signifies something to it: food, flight, reproduction—or, for that matter, despair." (1996: vii).

It may be misleading to attribute mind or thought to bacteria, protists, fungi, and plants, as well as to most invertebrate animals. But birds and mammals are certainly endowed with a mind, interpreted as the functioning of its nervous system. The world is thus meaningful to them as well as to humans. It is, however, rather misleading to view such "signs of life" as constituting an autonomous system or "semiosphere" that is completely separate from living organisms, the biosphere, and its physical setting. To suggest that "life is based *entirely* on semiosis" (Hoffmeyer 1996: 24) is perhaps an exaggeration.

But of course, for humans what is important is not only the "signs" experienced in our interactions with the material world, but also the linguistic "symbols" that enable humans to communicate with each other—what Hoffmeyer describes as "horizontal semiosis" (1996: 59).

Long ago Charles Peirce, in his theory of semiotics, made his famous distinction between three types of signs. He identified, namely, the *index*, in which signs are a part or an effect of that which they signify (such as a deer track, or a snake being a sign of danger for a vervet monkey); the *icon*, in which the relationship is based on resemblance (such as an image or portrait); and, finally, what Peirce describes as a *symbol*, in which the relationship between the sign and the signified is not so much arbitrary, as conventional, within a particular social context. Thus, "All words, sentences, books and other conventional signs are symbols." (Buchler 1955: 102-112). Unlike Ferdinand de Saussure, the pragmatist Peirce situated symbols—human language—firmly within a social and material context.

It is of interest that Hoffmeyer declined to engage with Saussure's semiology, which had, of course, a profound impact and influence on such scholars as Althusser, Lacan, Levi-Strauss, and Jacques Derrida.

A symbol (from the Greek, *symbolon*), is always a perceptible entity, whether audible words (sounds), an object, a gesture, or visible marks or inscriptions on some material thing. It is a sign which indicates or *represents* something, usually described as the meaning of the symbol. Always conventional, and always implying an interpreter, the meaning of a symbol is usually defined in two interconnected ways: in terms of its *reference* to some concrete or material entity (denotation), and in terms of its *sense*, in designating some concept or idea (connotation). The conjoining of both the sense and reference of a symbol constitutes its meaning (Bunge 2003: 54-58).

It is quite misleading to imply that as words get their *meaning* from their alleged relations (of contrast) with other words, within a system of meaning structures—as Saussure emphasized—then there is no relationship between language and the material world, or that there are no meanings outside of language—as Derrida and his followers seem to imply. Words do not in any obvious sense refer to an object—as Wittgenstein implied—but they do in their *usage* within specific socio-historical contexts (Collier 1985; Tallis 2012: 117).

Concepts (meanings)—which essentially derive from our interactions with a material (and social) world—and *words* (signs)—as used in communications—are, as many scholars have stressed, distinct phenomena. The units of meaning are therefore concepts, but concepts in themselves are neither true or false: only pre-positions, whether expressed as statements or representations, are true or false (Mahner and Bunge 1997: 52).

Unlike postmodern (literary) and symbolic anthropologists, who tend to con-flate anthropology with ethnography, evolutionary biologists and realist anthropo-logists (and archaeologists) have always been fascinated with the *origins* of human language. Thus much as been written on what John Pfeiffer (1982) long ago de-scribed as the "creative explosion", the emergence of symbolic thought—with re-gard not only to the arts and religion, but also, in embryonic form, empirical science and philosophy. What seems evident is that around 75,000 years ago, or even perhaps earlier, human language emerged, along with many other emergent factors in the creation of modern humans. These factors are, namely: increased so-ciability and the development of what evolutionary psychologists term the "theory of mind" (the ability of humans to infer what other humans are thinking, feeling, and intending); the development of self-consciousness (Hallowell); increased manual dexterity and the use of tools, and the growing complexity of human inter-actions with nature-labour (Engels); an increase in the volume and complexity of the human brain, particularly the neo-cortex; and, finally, the discovery of fire and the formation of a symbiotic relationship between humans and wolves (dogs), which greatly enhanced human hunting skills and increased their intake of protein. All these factors were closely interrelated in the emergence among humans of en-hanced cognitive skills, symbolic language, and symbolic culture. It has been de-scribed as the "human revolution" (Marx and Engels 1968: 354-364; Hallowell 1974; Dunbar 2004; Barnard 2012, 2016; Knight 2016).

Two points may be made in relation to the present discussion.Firstly, human cognition (and mental life more generally) is intrinsically both neural—a function of the brain—and social, as anthropologists have long emphasized. Secondly, a distinction can be made between human culture (or symbolic thought) and lan-guage, even though they are intrinsically linked. As one scholar has expressed it, culture is "how we live"—human life-ways relating to the values, ideas, beliefs, and narrative expectations that are shared by a particular society or group of people—while language is "how we talk" (Everett 2012: 6). Language is thus a cultural tool, employed mainly by humans as a sophisticated form of communica-tion; as a mode of expression with respect to intentions, feelings, emotions, and thoughts; and, finally, as a way of initiating performative acts, as well as having a cognitive function in conveying information or knowledge about the natural and social worlds in which humans live. Language is multi-functional, and it is quite misleading to suggest that it is purely expressive or that it has no representational function (Deleuze). For language allows us to produce descriptions and representations (whether cultural worldviews or scientific theories) of the material world. But as I stressed above, representations are symbolic forms produced through empirical enquiry and reason; they do not simply "picture" the world (Wittgenstein), nor are they a "mirror" of nature (Rorty), as empiricist (naïve) realists long supposed.

The postmodern mantra that language is purely expressive and does not offer a representation of the world, implies a rather vacuous dichotomy. An expression like "the cat is on the mat" is a statement about the material world—a re-presentation (not a mirror image)—of our perceptions. Even the expression "I'm thirsty" implies a representation of one's own bodily state. To set up a dichotomy between expressive and representational modes of thought (or language) is therefore quite misleading.

A rejection of logocentrism, interpreted as implying that language is a mirror or reflection of things in the world by natural resemblance (Derrida 1976: ii)—does not imply a rejection of reference, representation, or truth as correspondence (Assiter 1996: 60-66). Thus, semantic theories of *meaning* ought not to be conflated with the notion of *truth* as correspondence.

One can therefore reject "objectivism" without this entailing a rejection of either a realist metaphysic or the correspondence theory of truth (Johnson 1987: 200-212). The suggestion that there is no "transcendental signified", no world to which language can have any reference, is pure linguistic idealism—a position Derrida, in contradictory fashion, always firmly denied. But this is how Iris Murdoch interprets Derrida's philosophy, and his mode of textualism, which, like Bunge, she describes as consisting largely of "dramatized half-truths and truisms" (1992: 185-188; c.f. Hacking 2002: 67). The same can be said with respect to much of Heidegger's philosophy.

If by the "metaphysics of presence" Derrida simply intended a critique of "naïve realism" (Gratton 2014: 207) he is not saying anything particularly new, fresh, or original. People as different as Roy Wood Sellars, Lewis Mumford, and Ruth Benedict were emphasizing the social mediation of perception (and knowledge) several decades before Derrida, doing so with much less pretension and scholastic wordplay.

Knowledge as truth and as the representation of some given object of study—such as human-animal relations in Malawi (Morris 1998) or local cultural schemas and practices—entails, of course, a re-presentation, the making present to a reader of what is actually absent. This involves the unique gift of the human mind—imagination (Arendt 1978: 76). The notion that scientific thought does not involve the imagination is one of the popular misconceptions of science—but science also involves the critical testing of evidence for a particular theory (representation) which thus makes it quite distinct from poetry and literature (Medawar 1982).

When Kirsten Hastrup, for example, suggests that anthropology consists "not of representations but of propositions about reality" (1995: 45), though apparently acknowledging a realist perspective, she misleadingly defines the first term—representation—as implying an isomorphic or "mirror" image of the social world. This leads her to repudiate the "realist" monograph, which she caricatures as representing societies as "timeless, island-like entities", and to conceive ethnography as a "creative process of evocation or re-enactment" (1995: 21). If by "evocation" Hastrup means the representation of some given social reality that can be critically scrutinized by others, particularly the members of the community involved, then this seems barely different from what social scientists and ethnographers have been doing for a long time. If, however, by "evocation" she means the disclosure of truth in the manner of Heidegger (that is, to engage in hermeneutics), or as a poetic meditation that describes and explains "nothing" and is "beyond the truth"—in the fashion of Stephen Tyler (1991)—then such a strategy can hardly provide us with valid knowledge. Hastrup's enlivening study is bedevilled throughout—as with many hermeneutic scholars—by the conflation of realism and the correspondence theory of truth with positivism and an "objectivist" or reflective theory of meaning, misleadingly assuming—like Deleuze and Rorty—that "representation" necessarily entails an isomorphic, unmediated relationship between a theoretical account and the social world—as if such an account were indeed possible. But the search for truth through representation involves neither objectivism (or naïve realism), nor a poetic evocation of "nothing".

The repudiation of realism and the classical (correspondence) theory of truth by hermeneutic scholars and postmodern anthropologists seems to be largely due to the baneful influence of Martin Heidegger on the social sciences. The writings of this reactionary thinker, which I have discussed at length elsewhere (Morris 2014A: 579-594), now have the status of sacred texts among many admiring acolytes (Dallmayr 1993; Foltz 1995). Along with Nietzsche, he was certainly a key figure in the rise of postmodernism, though both scholars would probably repudiate the cultural idealism—social constructionism—and relativism of their purported followers.

Heidegger was not an ecological thinker (as frequently depicted), but rather a scholar whose worldview embraced the idea of a "four-fold" nature of being: earth, sky, mortals (humans), and the gods, the divinities being the messengers of the "godhead" (1978: 351). Nature for Heidegger had no history, and there was an absolute "abyss", he argued, separating humans from other earthly beings. He does not appear to have read Darwin! Like the latter-day positivist Bruno Latour, Heidegger was an empirical realist (phenomenalist), suggesting that the "world is only if and as long as Dasein (humans) exist" (1988: 241). Most of Heidegger's ideas are, in fact, a re-cycling of Aristotle's philosophy.

As an empirical realist, Heidegger was seeking a primordial or poetic relationship with things (beings) in the world, and he never denied, though he continually disparaged, the classical theory of truth as correspondence. He rather sought to express—often in mind-boggling abstractions that Bunge constantly mocked (1999: 230)—a primordial sense of truth. This was truth as *alethia* (derived from a reading of Aristotle): truth as disclosure or unconcealment, which he saw as prior to truth as "agreement of intellect and the thing". Thus, the phenomenological approach to both nature and (human) history promises, he writes, "to disclose reality precisely as it shows itself before scientific inquiry as the reality which is given to it" (1978: 176).

Heidegger, as his student and friend Hannah Arendt insisted (1978: 19), used the terms "meaning" and "truth" rather loosely and interchangeably. Thus what Heidegger—like Husserl—seemed to be suggesting is the quite uncontentious notion that before one can ascertain the truth (as correspondence) of a representation, or *explain* a particular phenomenon, one must first disclose its meaning, or ascertain a "primordial" understanding of it as a phenomenon.

Heidegger largely seems to be insisting—in endless repetition (1962: 257-273, 1978: 115-138)—on what has largely been taken for granted by generations of social scientists, or least those who have distanced themselves from crude positivism, namely, that one must engage in interpretive understanding or hermeneutics (meaning, *verstehen*) before one can explicate social and cultural phenomena through causal mechanisms (morphogenetic processes), or through contextual or historical explanations.

Empirical knowledge, and truth through representations, does not imply some absolutist metaphysics but a realist ontology; nor does it imply that what is being portrayed are some transcendental verities; still less does it entail a Faustian perspective, and the control and domination of nature (see Tyler 1991). What knowledge as representation does, however, is to make *explicit* what in fact is being affirmed (truths about the natural and social worlds), acknowledging that all truth is intersubjective and thus open to critical scrutiny and possible refutation by other scholars (unlike truths which are apparently disclosed through evocation or mystical "revelation", and which, we are told, have no reference at all to any world outside the text—or "discourses" or "theory"). With regard to anthropology, this affirmation of truth is particularly important, for ethnographic accounts should be open to scrutiny by the people whose social life and culture is being described and explicated. All knowledge is intersubjective, as even positivists long ago recognized, and thus only approximates to the truth (Feigl 1953). Anthropological knowledge expressed in the form of representations always implies a *perspective* on the world, always relates only to an *aspect* of the material and social world, and always entails painstakingly *empirical* research.

The notion that an earlier generation of anthropologists was enwrapped in a "visualist" metaphor—the postmodernist denigration of vision is quite fallacious!—or that these anthropologists lacked any imagination and saw themselves as completely detached observers, neither participating in the social life of a community nor being engaged in dialogue, is quite misleading. Such a crude positivistic or behaviourist stance would never have yielded the rich ethnographies of Malinowski, Boas, Evan-Pritchard, Meyer Fortes, Audrey Richards, and Irving Hallowell—to name but a few.

The postmodernist critique of anthropology, which is linked with its misplaced condemnation of the Enlightenment, and which was well exemplified by Johann Fabian's *Time and the Other* (1983), entailed a complete "misrepresentation" of anthropology which at times bordered on caricature. This critique consisted of three strands, namely, 1) that anthropology treats the people it studies in terms of a "radical alterity", that is, anthropologists distanced themselves from the people they studied, failing to recognize any common humanity; 2) that anthropology ignored history, viewing other cultures as "timeless" and unchanging; and, finally, 3) that anthropology treated such cultures as isolated entities, as hermeneutically sealed from other cultures, and from the wider capitalist world system.

In a seminal essay, Herbert Lewis (2014: 1-26) has countered in some detail this gross "mis-representation" of anthropology by its postmodern critics (Fabian 1983; Abu-Lughod 1991; Turner 1993; Keesing 1994). He thus emphasizes that anthropology has not only always recognized the importance of cultural difference, but also fervently believed, ever since Edward Tylor, in a common humanity and the importance of cultural universals.

Likewise, the idea that anthropology lacked a historical perspective is, Lewis contends, completely untrue: not only with respect to American scholars who were situated in the "historical" school that stemmed from Franz Boas (Alfred Kroeber, Robert Lowie, Alexander Goldenweiser, and Melville Herskovits, for example), but also with regard to British anthropologists. Although often described as structural-functionalists, Isaac Schapera, John Barres, Ian Cunnison, and Ioan Lewis all published historical studies (see I.M. Lewis 1999: 1-25).

As far as anthropologists treating non-Western people as social isolates, this too, according to Herbert Lewis, is a complete misrepresentation, by the postmodernists, of the work of an earlier generation of anthropologists. For anthropology has long been concerned with social change, acculturation (culture contact), and with the impact of the fur trade, mining, money and the market economy, and of European colonialism on the people of what was once described as the Third World. (H. Lewis 2014: 17-21)

Of interest is that even in 1936, Monica Hunter (Wilson) titled her ethnography of the Pondo of South Africa *Reaction to Conquest*, obviously not treating these people as some timeless, isolated culture.

Hence Herbert Lewis sums up the real nature of anthropology—not the caricature depicted by its postmodern critics—when he writes: "As a field it has produced a vast storehouse of knowledge about the peoples of the world, with neither the intent nor the result of conquering or dominating them—a rich literature of concern for both the universals, the things that all humans share, and the differences among us, and how these might be explained." (H. Lewis 2014: 21).

It is important to recognize that anthropology has always had a world-historical focus, that it has never been concerned purely with studying "exotic cultures", and that there has long been a tradition of "anthropology at home". Moreover, it is worth noting that many early pioneer anthropologists—though now largely forgotten—came from non-western backgrounds; for example, M.N. Srinivas and A. Aiyappan (India), Jomo Kenyatta (Kenya), and Fei Xiao-Xung (China).

To sum up the chapter, we may stress that there is an intrinsic connection between the correspondence theory of truth—as representation—and a realist ontology, and together they offer a viable alternative to both naïve objectivism (positivism), and cultural idealism (social constructionism), as well as those scholars who appear to see nothing between absolute truth and subjectivism. The latter—subjectivism—in contrast to realism, entails the view that the world of nature, far from existing independently, is the creation or "construction" of humans—either as a rational ego or a cultural subject. But, as indicated earlier, this is an entirely anthropocentric perspective (Bunge 1996: 330; Sayer 2000: 69; for further affirmations of the correspondence theory of truth see Sayers 1985: 177; Searle 1999: 13).

Having outlined and defended against postmodernism the concepts of representation and truth (as a correspondence to reality), I turn now to elucidating the parameters of a historical anthropology. I shall do so in terms of three conceptual dualities that need to be both acknowledged and dialectically combined, namely the dualities of individualism/holism, structure/agency, and hermeneutics/science.

Chapter Eleven

The Dialectics of Social Life

In part two of this manifesto, I outlined the various forms of materialism that constitute the metaphysical basis or background to my advocacy of anthropology as a historical social science. The materialism invoked was one that was realist, historical (evolutionary), dialectical, systemic, and emergent. Having affirmed the salience and importance of the concepts of both truth and representation, I now turn to specifically discussing the dialectics of social life. This chapter is primarily concerned with overcoming two dichotomies that are pervasive in the philosophy of the social sciences, namely, individualism versus holism (addressed in the first section) and social structure versus human agency. I stress the importance of developing a dialectical social theory.

Individualism and Holism

In the first chapter of this manifesto, I indicated that for many scholars—Erich Fromm (1964: 116) in particular—there is an essential "tension", or "contradiction", at the heart of human life. As Kenan Malik expressed it: humans are "both inside nature and outside it. The peculiar position that human beings occupy in the natural order means that we require special intellectual tools to understand ourselves." (2000: 339).

There is, then, an inherent duality in social existence, in that humans are an intrinsic *part* of nature, while at the same time—through our self-conscious experience, our intense sociality, and our symbolic culture—we are also in a sense *separate* from nature. Lewis Mumford therefore writes of humans living simultaneously in "two worlds". The first is the material world; the other is what many scholars, following Cicero, have described as our "second nature"—the social and symbolic life of humans that is "within" first nature, and which makes humans a unique species (Mumford 1951: 48; Dubos 1973: 102; Bookchin 1989: 25).

Many of the debates and harsh polemics within both philosophy and the social sciences therefore stem from the fact that humans have a "two-fold life", as Mumford (1951: 48) again expressed it—that we are both natural and social beings. Thus we have contrasting emphases or perspectives: materialism (realism) versus idealism (discussed in earlier chapters), individualism versus holism, naturalism (science) versus humanism (hermeneutics) (Bunge 1998: 4-60). These dichotomies are central to anthropological understanding and are each discussed below.

As many scholars have acknowledged; human life therefore involves a central "paradox" (as Husserl describes it). There is an inherent duality in human existence, in that we are contemplative beings who, through conscious experience and symbolic culture, see ourselves as separate from the material world, while at the same time we are active participants within this world as natural (biological) beings. But our conscious awareness of the surrounding lifeworld (*lebenswelt*), essentially expressed in visual metaphors, does not in the least imply that vision, insight, and observation entail complete detachment or distance from the material world. Hence the paradox of humans as simultaneously "world constituting" (giving meaning to) and causally related to the world (Husserl 1970: 262). An overemphasis on symbolism and subjectivity (as with existentialists, phenomenologists, and cultural idealists), or on the natural or objective dimension to human life (as with evolutionary psychologists, sociobiologists, or reductive materialists) are, however, both limiting, one-sided perspectives. A dialectical approach to the human "paradox" is, I think, therefore essential.

In similar fashion, humans are both personal and social beings, ontologically distinct from social life, but at the same time constituted through it. Social reality—or what Richard Jenkins describes as the "human world" (2002: 64-67)—unlike the material world, is not self-subsistent but is dependent on human activity. At the same time, like nature, it is transformed and shaped by human agency. There is then an "ambivalence", as Margaret Archer (1995: 2) writes about social reality: social and cultural structures are dependent on human activity, while at the same time, through social practices, these structures have a constraining and determining influence over us as individuals. This was well expressed by Karl Marx who wrote that "Men make their own history, but they do not make it just as the please; they do not make it under circumstances chosen by themselves, but under circumstances directly encountered, given and transmitted from the past. The tradition of all the dead generations weighs like a nightmare on the brain of the living" (Marx and Engels 1968: 96).

Human beings thus, through social praxis, create social systems and normative structures, along with cultural schemas (symbolic forms), which, as emergent entities (or processes) in turn constrain and condition individual consciousness and behaviour. The history of the social sciences has, therefore, long seen an ongoing debate—bordering on a dispute—between two quite distinct ontologies or approaches to social life. Indeed, Alan Dawe (1979) interpreted the "persistent tension" between the two approaches as an immediate expression of the inherent "dualism of social experience". The first approach has been variously described as holism, collectivism, or "social systems" theory; the second approach as atomism, methodological individualism, or "social action" theory (Cohen 1968; Hollis 1994: 5-12; Archer 1995: 35-54; Bunge 1996: 241-263). I shall discuss each approach in turn.

The first approach, holism, puts a decisive emphasis on society, social structures, or human symbolic culture as the fundamental reality. It treats human agency and individual consciousness as ephemeral. Durkheim's sociology—along with that of Talcott Parsons (1937)—and Marx's famous preface to *A Contribution to the Critique of Political Economy* (1989) are often taken as exemplifying this approach. For in an oft-quoted phrase Marx wrote: "It is not the consciousness of man that determines their being, but on the contrary, their social being that determines their consciousness" (Marx and Engels 1968: 181).

This "top-down" approach, as a form of social determinism or what Archer (1995) describes as "downward conflation", essentially presents humans as *Homo sociologicus* (or, alternatively, *Homo symbolicus*), and tends to dissolve personal identity and human agency into social structures (relations) or cultural configurations. As I have discussed more fully elsewhere (2014A), this approach can be viewed as characteristic of structural-functionalism (Talcott Parsons and Radcliffe-Brown), mechanistic versions of Marxism (Althusser), structuralism (Levi-Strauss), and much symbolic (or cultural) anthropology (Geertz), especially the culture-and-personality school. This latter current of thought tended to view the human subject as simply a "microcosm" of cultural configurations. Indeed, anthropologists such as Ashley Montagu and Ruth Benedict played an important role in what Kenan Malik describes as *Unesco Man* at the end of the Second World War. This emphasized that humans were fundamentally *cultural* beings and that a plasticity of mind was a species-characteristic of *Homo sapiens* (2000: 134-139). As with the later postmodernists, this approach suggested that there was no such thing as human nature, and that the human mind was simply a "blank slate" (Pinker 2002). In the process, human agency, as expressed in social practices, tended therefore to be completely downplayed.

On both ontological and epistemological grounds, many scholars have been critical of holism (or organicism) as an approach to social life. It has been suggested that it tends to reify "society" or "culture", or at least view them as totalities that completely transcend the human individual, as well as implying that social facts (or individual actions) are explicable *solely* in terms of these supra-individual entities. But as Mario Bunge writes, social systems (or cultures) do not have a life of their own, for there is no social system without components (individual human beings). Social systems, he writes, "are nothing but systems of inter-connected persons together with their artefacts" (1996: 260-261). Bunge even suggests that Marx, in over-emphasizing the social matrix of individual action, "lost sight of the individual" (1998: 64).

Such a holistic approach was given a new lease of life by the postmodernists, discussed in the opening chapter, whose critique of the transcendental subject of Cartesian metaphysics seems to have gone to extremes in eradicating human agency from the analysis. The human person was either erased entirely—the "end of man" syndrome—or the human subject was seen simply as an "effect" or a "construction" of power, ideology, discourses, or language. As one anthropologist put it, blindly following Foucault: "We as subjects are constructed in discourses attached to power" (Abu-Lughod 1991: 158; see also Coward and Ellis 1977: 74; Flax 1990: 231).

Although acknowledging the reality of social systems or cultural representations, and their constraining influence, via social praxis, on human life, it is especially important to affirm the salience of human identity and of individual agency. But the human person must be conceptualized not as a transcendental epistemic subject, but as an embodied being, embedded in both an ecological and a social context (Benton 1993: 103; Morris 1994: 10-15).

I have elsewhere discussed at length the varied conceptions of the human subject within the Western intellectual tradition (2014A). Given the human paradox, it is not surprising that many approaches seem to gravitate to extremes. On the one hand, there are those scholars—existentialists, cultural anthropologists, and postmodernists—who fervently deny that there is such a thing as "human nature", and often view the subject as simply an "effect" or a "microcosm" of culture, ideology, power, or language. This approach is untenable as it completely oblates human biology and individual agency.

At the other extreme, there are those scholars—sociobiologists, evolutionary psychologists, economists, and rational choice theorists—who firmly believe in the existence of a universal human nature. But there has been a notable tendency, particularly by sociobiologists, to depict human nature in terms of the rational ego: aggressive, competitive, possessive, territorial, and xenophobic. This portrayal of human nature reflects the ethos of laissez-faire capitalism and is equally untenable. It lacks any cultural sensibility or any recognition of the uniqueness and agency of the individual person.

Avoiding those two extremes, many scholars have emphasized that humans are fundamentally *both* biological (natural) and social (historical) beings; this dualistic conception of human subjectivity was well expressed by Emile Durkheim as *Homo duplex* (1912: 16; Morris 2014A: 210-211).

But drawing on a wide range of scholars, as varied as Marcel Mauss, Clyde Kluckhohn, Irving Hallowell, Erik Erikson, and C. Wright Mills, I have contended that the human subject is best understood in terms of a triadic ontology. The human subject has therefore to be conceptualized in relation to three interconnected aspects or components. These are the human subject as a *species-being*, characterized by universal biopsychological dispositions, self-consciousness, and sociality; as a unique individual *self*, embodied and embedded within a specific historical and ecological context; and, finally, as a social being or *person* enacting multiple social identities or subjectivities. The human subject therefore not only has a *human* identity and a *self* identity, but also various *social* identities, relating to the social-structural aspects of human life (Jenkins 2008; Morris 2014A: 760-769).

Humanism, as a belief in the power of human agency in history, must therefore be affirmed, and should not be equated with either possessive individualism (as reflected in political liberalism and classical economics) or with Cartesian metaphysics, but rather with the aforementioned triadic conception of human subjectivity.

Even less should humanism be equated with a socialized version of Christian faith, one that posits a dualistic metaphysic that not only implies a radical "gulf" between humans and other life-forms, but suggests that humans have been given dominion over the earth, expressed in the technological mastery of nature

(Braidotti 2013). This is how scholars like Ehrenfeld (1978) and Gray (2002) have defined humanism, painting a rather misanthropic portrait of the human subject as an inherently destructive and predatory being who is thus in need of salvation or redemption via some religious faith or mysticism. In contrast, the Enlightenment conception of humanism, discussed in chapter three, affirms the crucial importance of human agency and evolutionary naturalism as an ontology, while extoling a form of ethical naturalism and emphasizing free enquiry and the importance of reason and empirical science (Kurtz 1983: 39-47; Bookchin 1995B: 12-14; Bunge 2001A: 14-15).

It is worth emphasizing again that a human perspective on the world—expressed philosophically as humanism or universalism—is recognized by all human societies, often explicitly. African people in Malawi, for example, whatever their ethnic affiliation, often emphasize in everyday life the importance of *umanthu* (humanness) as an ethical principle in guiding behaviour (Morris 2000: 52). To interpret such a humanist perspective—distinct from one that is purely subjective or culturally specific—as implying an anthropocentric ethic, and thus as sanctifying human sovereignty and domination over nature, is misleading if not quite facile (c.f. Ingold 2013B).

The second approach, that of *individualism*, rests upon the empiricist conviction that the ultimate constituents of social reality are human individuals or people. It thus treats social structures and cultures as epiphenomena. The prototype statement comes from John Stuart Mill, who wrote: "The laws of the phenomena of society are, and can be, *nothing but* the actions and passions of human beings united together in the social state. Men, however, in a state of society are still men; for their actions and passions are obedient to the laws of individual human nature." (1987: 65). Or, as he also put it: all the phenomena of society may be "resolved into the laws of the nature of individual humans—the phenomena of human nature" (op. cit. 63).

This became the fundamental stance of many methodological individualists and empiricists, who affirm that the "ultimate constituents of the social world are individual people who act more or less appropriately in the light of their predispositions and understanding of their situation" (Watkins 1968: 270). Marx and Engels are also linked to this approach, for they wrote (1845): "History does nothing, it does *not* possess immense riches, it does *not* fight battles. It is men, real living men, who do all this, who possess things and fight battles.... History is nothing but the activity of men in pursuit of their ends" (Marx and Engels 1956: 93; Flew 1985: 61-66).

As distinct from the "view from on high", individualism or social action theory is a "bottom-up" approach, a form of "upward conflation", that reduces social structures and cultural schemas to the social actions and dispositions of individual humans. Individualism focusses specifically on individuals, and either denies the existence of social bonds or social systems or holds that they are reducible, as Mill implied, to individuals and their immediate actions. It is a view, as Bunge suggests, that goes back to Hobbes and it is shared by all utilitarian philosophers (Bunge 1996: 243).

This approach has not only been associated with many methodological individualists—for example, Karl Popper, Friedrich Hayek, George Homans, and Ayn Rand (Lukes 1968), but also with the "rational agent" of classical economics and rational-choice theory. Rand, of course, was the intellectual guru of the neoliberal and conservative Margaret Thatcher, who famously declared that "There is no such thing as society. There are only individual men and women and their families." Of course, she never doubted the existence of a coercive nation-state and the capitalist economy (Kingdom 1992; Outhwaite 2006: 17). It even became fashionable amongst postmodern anthropologists, like Marilyn Strathern, to assert that the concept of "society" was theoretically "obsolete" (Ingold 1990).

But the individualist (or subjectivist) approach has also been expressed in various forms of interpretive sociology—particularly ethnomethodology (Garfinkel 1967) and symbolic interactionism, as well as in the latest avatar of ethnomethodology, actor-network theory. The acronym ANT is often used as a marketing ploy! The latter approach, inspired by Gabriel Tarde, adopts the positivistic approach of methodological individualism, thus repudiating both socio-historical explanations, and views sociology as a kind of "interpsychology" (Latour 2005: 13; for an amusing rebuttal of this approach see Ingold 2011: 89-97).

The shortcomings of individualism to the understanding of social life have been highlighted by several scholars, at least with regard to its more radical versions. Mario Bunge, for example, stressed two important limitations to this current of thought, although rational choice theory is now regarded as the most refined and influential paradigm in the social sciences (Steel and Guala 2011: 211). The first is that, in stressing the absolute primacy of the rational individual as well as the legitimacy of the relentless pursuit of individual interests—it offers a naïve utilitarian view of human agency—the individualist approach leads to a rather ahistoric and impoverished conception of the human subject. Emphasizing (instrumental) rationality over sociality, it also neglects the importance of the human emotions. Secondly, it tends to overlook or downplay the social context of human agency, for individual intentions, expectations, and desires are to a large degree motivated and shaped by social structures and social (and cultural) circumstances. As a form of positivism, individualism thus neglects socio-historical factors and causal (social) explanations (Bunge 1999: 48; Archer 2000: 51-74).

It may be noted that the individualist approach, and the rational choice theory in particular, seem to coalesce rather well with the pre-occupations of management consultants, studies of marketing, economic organizations, and the ideology of capitalism more generally. In fact, the theory is often presented as a "blanket endorsement" of free-market capitalism, and of the economic man—*Homo economicus* (Bunge 1996: 248-252, 1999: 87-101; for useful accounts of rational choice theory see Coleman 1990; Baert and Carreira Da Silva 2010: 125-154; Pettit 2011).

Going beyond the limitations of both holism and individualism in the understanding of human social life leads us inevitably to exploring the relationship between individual agency and social structures—the enduring relationships within social systems. To this issue I now turn.

Structure and Agency: A Dialectical Synthesis

There have been many attempts to transcend the "duality" of structure and agency: Anthony Giddens' (1954) structuralist theory and "ontology of praxis", Michel Foucault's (1980) genealogical approach and emphasis on "power-knowledge", Norbert Elias' (1978) theorizing around the concept of "figuration", and Michael Jackson's (1989) radical empiricism may all be taken as examples. But what these approaches have in common and tend to entail is a *conflation* of the central dynamic or dialectic between individual human agency and social structures. It is a dialectic that is oblated with an undue emphasis on, respectively, social practices, discourses of power, social configurations, or lived experience (Layder 1994: 94-149; Archer 1995: 101).

What many scholars have therefore suggested is the need for a critical realism, a systemic or relational (dialectical) approach that acknowledges *both* human agency and social structures (or cultural representations) as distinct levels of reality, each having emergent properties that are irreducible to one another. As Margaret Archer writes in her advocacy of a morpho-genetic approach to sociology, social structures as emergent entities "are not only irreducible to people, they pre-exist them, and people are not puppets of structures because they have their own emergent properties which means they either reproduce or transform social structure rather than creating them" (1995: 71).

Although Archer refers to this approach as "analytic dualism", she advocates a dialectical, not dualistic approach, an analysis that "links" people to their social context. The interplay between structural and social agency is historical, in that the "morphogenetic cycle" involves social structures which are in a sense prior to, and condition, social interaction and agency, which in turn transform or reproduce the social structures. She repudiates any recourse to metaphor in describing social reality or the human life-world, for a society is not like a language or text: it is not a mechanism with fixed parts, nor is it a cybermatic (homeostatic) system, nor yet is it a kind of theatrical performance, or a piece of textile fabric (a favourite metaphor among cultural anthropologists). Society, rather, is only itself as a material entity—ordered, enduring, open, processual, and peopled (Archer 1995: 166-194). The idea, expressed by some postmodernist scholars, that an earlier generation of anthropologists viewed societies as *closed*(physical) systems is quite fallacious (I.M. Lewis 1999: 1-25).

Archer therefore emphasizes the "duality" of social existence and suggests that the basic task of the social sciences is to conceptualize how ordered social formations have their genesis in human agency, while humans, as social beings, in turn have their genesis within social forms or social systems (1995: 167). It has been suggested, however, that the exact "linkage" or interactive processes between human agency and social structures tend to be rather neglected in Archer's conceptual framework (Mouzelis 2008: 210), and the historical and ecological contexts of social systems also seem to be rather neglected.

Both the classical approaches to the understanding of human social life—individualism and holism—are therefore one-sided and inadequate. An emphasis on methodological individualism reduces social life to human dispositions and actions—and humans are invariably depicted as Hobbesian individuals—rational, competitive, acquisitive, self-interested, autonomous, and maximisers of their own utility. On the other hand, in treating collective phenomena—social systems, power, cultural representations, discourses—*sui generis,* as the fundamental reality, with regard to which humans are simply an "effect" or "product", entails a *reification* of social phenomena and is equally untenable. As Erich Fromm put it, a human being is not a "blank sheet of paper on which culture can write its text" (1949: 23); humans are physical beings with natural capacities and powers, are charged with energy, and are structured in specific ways. Humans and their self-identity are not then simply an "effect", or a "construct" of social structures, power, cultural configurations (ideology), or discourses.

What is therefore needed is a theoretical perspective that *combines* both approaches. It is a perspective initiated by Karl Marx who, significantly, is seen as exemplifying both approaches. For Marx indeed emphasized both human agency and the ontological reality of social institutions and ideologies as emergent phenomena. What then is required is a *dialectical* approach that combines individualism and holism, as well as humanism and structuralism, providing a "linkage" theory or a "relational model" of social life. This model suggests that social being is constituted through social praxis, for our social acts presuppose the existence of social systems and shared cultural schemas, values, and beliefs. Yet, at the same time, humans are seen as ontologically independent of social relations. As personal beings have agency, self-consciousness, and self-identity (Harre 1983), there is then a need—contra postmodernism, and the alleged "dissolution" of the human subject or the self (c.f. Gergen 1991: 6-7; Archer 2000: 19)—to reclaim human agency for social analysis, without this implying a relapse into transcendental subjectivity or the acceptance of the bourgeois, abstract individual espoused by the economists and rational choice theorists. The emphasis on social structures (systems) indicates the way in which cultural schemas (expressed in language and ritual) and social institutions come, through social practices, to shape (and modify) human consciousness and behaviour. The emphasis on human agency equally highlights the degree to which humans, as embodied selves, change social structures and cultural frameworks (Collier 1994: 140-141; Layder 1994: 209-210).

Many scholars throughout the history of the social sciences have of course attempted such a synthetic or dialectical analysis. Three recent scholars who exemplify such a theoretical attempt to combine individualism (subjectivism) and holism (objectivism) are perhaps worth briefly mentioning in this present context, namely, Pierre Bourdieu, Mario Bunge, and Lucien Goldman. I shall discuss each scholar in turn.

A unique inter-disciplinary scholar who combined anthropology, sociology, and philosophy, Pierre Bourdieu stood firmly, in important respects, within the social scientific tradition of Marx, Durkheim, and Weber. His reflexive sociology, however, had many affinities to John Dewey's philosophy of pragmatism—a form of empirical naturalism. For like Dewey, Bourdieu was deeply suspicious, if not hostile, to the deep-seated intellectualism that was characteristic of Western philosophy—what Dewey described as the "spectator" theory of knowledge (Dewey 1916A: 393). Thus, like Dewey, Bourdieu strongly emphasizes a "logic of practice", granting a central role to the notion of habit, understood as an active and creative relation to the world. Finally, echoing Dewey, Bourdieu expressed a strong opposition to all the conceptual dualisms that stemmed initially from Cartesian philosophy: between subject and object, spiritual and material, individual and society (Bourdieu and Wacquant 1992: 122). Thus at the very beginning of his seminal sociological text *The Logic of Practice*—written, again like Dewey, in the most dense, almost impenetrable prose-style—Bourdieu writes: "Of all the oppositions that artificially divide social science, the most fundamental, and the most ruinous, is the one that is set-up between subjectivism and objectivism." (1990: 25).

Both these forms of knowledge had their exemplars in the intellectual milieu in which Bourdieu worked. Subjectivism was reflected in Sartre's existentialism, Husserl's phenomenology, and rational choice theory; objectivism, or what Bourdieu tended to describe as "social physics", was expressed in Levi-Strauss's structuralist anthropology, positivistic sociology, and the structural Marxism of Althusser. Bourdieu aimed to develop a reflective sociology that would transcend the subjectivist/objectivist dichotomy, while at the same time "preserving the gains from each of them" (1990: 25). To this end, Bourdieu developed two key concepts, that of "habitus" and "cultural field", both of which have given rise to a welter of critical commentary. By *habitus* Bourdieu intended "socialized-subjectivity", the system of structural, durable dispositions, constituted in social practice, which were embodied in individuals. They result, he suggested, from the "institution of social in the body (or in biological individuals)" (Bourdieu and Wacquant 1992: 126-127). In contrast, a cultural *field* is defined by Bourdieu as a configuration on the "structure of objective relations between positions objectively defined" and which "guide the strategies of the occupants of these positions, within a specific distribution of power and various forms of capital (economic, social, cultural, symbolic)" (Jenkins 1992: 85; Bourdieu and Wacquant 1992: 101).

The central aim or theme of Bourdieu's work was, therefore, an attempt to understand the complex interrelationship between subjectivity—the lived experience of the individual subject (emphasized by existentialism and phenomenological sociology)— and the "objective" social world (emphasized by classical Durkheimian sociologists and anthropologists like Levi-Strauss). His project of "genetic structuralism" was thus an attempt to understand "how 'objective' supra-individual social reality (culture and institutional social structure) and the internalized "subjective" mental worlds of individuals as cultural beings and social actors are inextricably bound together" (Jenkins 1992: 19).

Critical of both structuralism and rational choice theory as ahistorical perspectives, Bourdieu emphasized that the relationship between the human individual and habitus—as a product of social practice—and a specific cultural field is always historical and subject to constant change (1994: 7). Therefore, Bourdieu posited an essential *dialectical* relationship between human agency and social structures, for habitus "contributes to constituting the field as a meaningful world, a world endowed with sense and value", while at the same time, the cultural field structures the "habitus" of individuals, in a relation of conditioning. Social reality therefore exists both "outside and inside of agents" (Bourdieu and Wacquant 1992: 127).

Bourdieu's essential thesis can be summed up as follows: "Social structures, via various socialization processes, are internalized and become dispositions (habitus) and dispositions lead to practices which in turn, reproduce social structures." (Mouzelis 2008: 130). But several scholars have suggested that Bourdieu's emphasis on habitus as socialized subjectivity tends, in Durkheimian fashion, to downplay the importance of creativity, reflectivity, and the rational strategies of individual human subjects (Bunge 1996; 259; Mouzelis 2008: 137).

I have above, in chapters eight and nine, outlined Mario Bunge's hylorealist perspective and his systemic philosophy (emergent materialism). Here I turn specifically to his social theory, for in similar fashion to Bourdieu, Bunge advocated a systemic approach to social life. This approach attempted to combine and transcend the dichotomy between individualism (subjectivism) and holism (objectivism), while recognizing, of course, that both these social ontologies contain an element or "nugget" of truth (1999: 45). As noted above, he offered some strident criticisms of both ontological individualism (ethnomethodology, phenomenological sociology, and rational choice theory), and holism (typified, for Bunge, by Durkheimian sociology), emphasizing their shortcomings and one-sidedness. As he put it: both ontologies have "the charm of simplicity", for individualists overlook the social constraints on the individual and present an "under-socialized" view of human subjectivity—especially the rational choice theorists (Granovetter 1985); while holists or collectivists underrate individual interests, initiatives, and agency, and present an "over-socialized" conception of the human subject (Wrong 1961; Bunge 1996: 257, 1998: 76). Given the complexity of the social world, Bunge contends we should thus steer clear of both extremes.

Bunge, therefore, comes to advocate a systemic approach that attempts a dialectical synthesis (not his phrase!) of individualism and holism, which suggests, in a nutshell, that human history "is made by individuals acting in and upon social systems that pre-exist and shape them" (1998: 279). Or, as he put it elsewhere, according to the systemic view: "Agency is both constrained and motivated by structure, and in turn the latter is maintained and altered by individual action. In other words, social mechanisms reside neither in persons nor in their environment—they are part of the processes that unfold in and among social systems." (1999B: 57).

For Bunge, social systems (for instance, family and kinship groups, schools, business firms, cultural institutions, prisons, tribal communities, or nation-states) are concrete systems or assemblages composed of real entities—people and the artefacts they use and communicate with—held together by social bonds, interactions, and enduring relations (social structure), that behave as a unity in some respects. Refusing to conflate social structure and social systems, and emphasizing that social relations, as emergent structures, are usually the property of a social system, Bunge also stressed that social systems are always embedded in nature. Such systems must therefore be conceptualized in terms of their components (active human beings and artefacts), their structure (as a collection of enduring social relations or interactions), and their environment (both social and natural). Bunge also argues that every human society (whether a tribal community or a nation-state) is composed of four sub-systems. They essentially express the four basic concerns or functions of human and non-human animal life. These are sex, mating, and reproduction (kinship); the provision of food and basic livelihood (warmth and shelter) through human interactions with nature (economics); issues relating to protection, the resolution of conflicts, and power (politics); and, finally, forms of communication (culture) (Bunge 1996: 270-271, 1998: 311).

These four functions relate to beetles as well as humans! This means, of course, that any explanation of social and cultural phenomena must involve a multiplicity of factors or causes, or what Bunge describes as the "five intertwined strands" of human history: ecological, biological, economic, political, and cultural. A systemic approach to human social life, for Bunge, is therefore a form of integral historiography (1998: 275).

It is worth noting, at this juncture, that recent scholarship has described the contemporary world of "globalization" as involving the "demise of the social"; that every thing "has become communication", or that the concept of "society" is now redundant, defunct, or outdated. It is therefore argued that "society" as a concept has been replaced by "politics" or "culture", or that contemporary social life consists exclusively of flows, networks, lines of flight, or "mobilities"—of people, capital, goods, new materials, ideas, images, and information (Urry 2003; Gane 2004: 91-104).

Culture and politics are simply aspects or dimensions of enduring social systems, and while not denying the complexity of so-called "globalization"—the expansion of industrial capitalism—and the importance of various informal networks, it must be recognized that "flows" or such networks always presuppose the existence and significance of social systems—whether local communities defending their lands against mining interests, N.G.O.s, the World Bank, transnational corporations, or nation-states (c.f. Outhwaite 2006).

In a paper written some thirty years ago (Morris 1985)—specifically on radical scholars Kardiner, Reich, Fromm, Laing, and Hallowell, who had attempted to bring a human/psychological dimension into social analysis—I critiqued the two extreme positions taken on the siting of human agency within the social sciences. The one extreme, as noted above, was to give human subjectivity and rational agency priority in social analysis. It was a strategy assumed by ego psychologists and methodological individualists. The other extreme was to expunge human agency from the analysis entirely—as implied by the culture theory of Leslie White, behaviourist psychologists, structural Marxists, many discourse theorists, and postmodernist scholars. For the latter, as I discussed above, the human subject or self is conceived simply as an effect of discourses or power-knowledge.

In the essay I drew attention to the writings of the important, but now largely forgotten Marxist literary scholar, Lucien Goldman. For in relation specifically to Althusser's structural Marxism, highly fashionable among anthropologists in the 1970s (see Morris 2014A: 634-649), Goldman rejected as untenable many of the radical dualisms that pervaded contemporary thought. He particularly criticized the phenomenologists and existentialists who gave analytical priority to the subject, or *cognito*, as well as the structuralism of Levi-Strauss and Althusser, which he argued had led to the complete "negation of the subject". As Levi-Strauss had famously put it: the "ultimate goal of the human sciences is not to constitute man but to dissolve him" (1966: 247). Goldman therefore argued, like Bourdieu and Bunge, that it was necessary to integrate the perspectives of phenomenology (subjectivism) and structuralism (objectivism), and to seek both the meaning and functionality of social structures. He thus suggested a *dialectical* approach that situated the "creative subject at the interior of social life" (Goldman 1977: 106).

It should be recognized, however, that the *ontological* distinction between social structures and human agency, and the respective emphasis given by the methodological individualists and the "collectivist" scholars—the latter stressing the priority of social structures, culture, or language (discourses)—is not coterminous with the *epistemological* division between interpretation (hermeneutics) and social science (naturalism). For example, Mill and Hayek, on the one hand, and Radcliffe-Brown and Leslie White, on the other, all advocate a scientific approach to social life as well as playing down hermeneutics. Yet they are on different sides of the fence in terms of their metaphysics. Indeed, White suggested that the most adequate scientific approach to culture should proceed as if "human beings did not exist" (1949: 41). Likewise, Durkheimian sociology, the structuralism of Levi-Strauss, structural Marxism, and postmodern anthropology have much in common in their tendency to repudiate human agency: but postmodernism (Tyler) puts a crucial emphasis on hermeneutics in an extreme form—textualism—and disavows both science and empirical social science. We may express these relationships schematically as follows:

	INDIVIDUALISM	HOLISM
HERMENEUTICS	Existentialism Phenomenology	Hermeneutics Cultural Anthropology Postmodernism (textualism)
SOCIAL SCIENCE	Psychoanalysis Rational Choice Theory Evolutionary Psychology	Structural Marxism Durkheimian Sociology Structural Anthropology

Having discussed the dialectics of social life in terms of overcoming and going beyond the contrasting social ontologies of individualism and holism, I turn in the final chapter of this manifesto to a discussion of epistemological issues. I discuss, in particular, the relationship between interpretive understanding (hermeneutics) and social science in the framing of anthropology as a historical social science.

Chapter Twelve

Anthropology - Science and Hermeneutics

In this final chapter I shall outline—but only outline—what I consider to be the essential tenets of anthropology as a historical (or humanistic) social science, advocating an approach to social and cultural life that combines hermeneutics and empirical science. Moreover, I will make the case for a social science that is both ecological and historical, in being grounded in a philosophy of historical or dialectical naturalism (materialism) that this manifesto has sought to extol.

Reflecting on the later writings of Evans-Pritchard, David Pocock suggested that his mentor had instigated a movement in social anthropology from "function to meaning" (1961: 76). This thesis completely overlooks the fact that one of the doyens of functionalist anthropology, Radcliffe-Brown, had long emphasized the importance of delineating the "meaning" of social and cultural phenomena, and that Malinowski—the other founding father of British social anthropology (Kuper 1973)—had also stressed the importance of understanding the "native's point of view" (Malinowski 1922: 25; Radcliffe-Brown 1922: ix).

Cultural anthropology in the United States, likewise, had emphasized the crucial importance of interpretive understanding long before Clifford Geertz (1973), as the work of Frank Cushing, Franz Boas, Paul Radin, and Gladys Reichard attest. It is also significant to note that Evans-Pritchard's important critique of Radcliffe-Brown's positivism, rather than advocating a radical hermeneutic or symbolic approach to anthropology, stressed the need to create a dialogue between anthropology and historical understanding (Evans-Pritchard 1962; Morris 1987: 188-189; I.M. Lewis 1999: 4-12).

In an important sense, then, hermeneutics and interpretive understanding have always been constituent parts of anthropology, especially in relation to ethnographic studies. It has been accepted as such even by cultural materialists such as Marvin Harris. What Harris challenged in the cultural idealist tradition of Geertz and David Schneider was not their hermeneutics and cognitivism per se, but rather their tendency to reduce all social life to semiotics, to treat all social facts as symbols, and to repudiate causal analysis and historical explanations—to deny, this is, that cultural representations, or symbolic forms, are explicable in terms of what Harris describes as "infra structural conditions" (1980: 258-282).

The so-called "interpretive turn" in the social sciences, which was reputed to have occurred around the middle of the last century, in the wake of the publication of Evans-Pritchard's seminal study *Nuer Religion* (1956), has to be understood largely as a reaction. It was a reaction to the positivist sociology specifically associated with Talcott Parsons (1937), Radcliffe-Brown's (1957) structural-functionalism (see Morris 2014A: 232-240), the scientism of much structural Marxism, and structuralist theory. Indeed, Josef Bleicher, writing in the early 1980s, had suggested that with the radical influence of the positivistic sciences, there had been a concomitant "atrophy" or "demise" of the "hermeneutic imagination" within the human sciences (1982: 1-2).

A decade later, it seemed the pendulum had swung to the other extreme. Rabinow and Sullivan's (1987) advocacy of the "interpretive turn" in the social sciences—along with postmodernism (discussed in the first chapter)—seemed to many to entail the complete repudiation of empirical social science. But Rabinow and Sullivan made it clear that the interpretive approach, with its focus on "cultural meaning", and with the emphasis that knowledge is "practical" and historically situated, did not imply a collapse into cultural relativism or the exultation of a romantic "subjectivism". They were also dismissive of Jacques Derrida's (1976) "textualism" in that it tended to completely oblate social praxis. But as with Geertz, their fundamental emphasis is on anthropology as a form of hermeneutics, involving the "interpretation of culture"—the latter being defined as a "web of signification", the shared meanings, practices, and symbols that constitute the human cultural world (1987: 7). They did not deny the persistence and theoretical fruitfulness of certain "explanatory schemas" in the social sciences, but these are never theorized.

A similar standpoint is taken by Charles Taylor (1985), who, in his advocacy of a "philosophical anthropology", stressed the crucial importance of a "hermeneutic component" in the human sciences. But what he critiques as "naturalism" is largely an outdated conception of positivistic science. This, for Taylor, involved: a mechanic paradigm, a "designative account" of meaning that implies an unmediated relationship between language and the world—again Wittgenstein's picture theory of language—and a "disengaged identity", a conception of the person as disembodied, atomistic, and as divorced from the social context (1985: 3-12). Such a "natural science" model (that is, the positivistic conception of science) has, of course, long been critiqued by the social sciences and by philosophy (see chapter eight), without this entailing the repudiation of either naturalism (materialism) or empirical science, still less the complete embrace of a reductive form of hermeneutics in the style of Heidegger. Taylor, however, seems to have little sympathy for those dismissive of the scientific outlook, and even less sympathy towards the obscurity and posturing that is reflected in Derrida's writings (1985: 10). Yet, although arguing that interpretation is essential to explanations in the human sciences, Taylor, like Rabinow and Sullivan, is silent when it comes to exploring what social "explanations" exactly entail. One can surely recognize that humans as organic beings are an intrinsic part of nature, and that social life is explicable by means of social science—naturalism—without this implying a reductive, mechanistic, and atomistic conception of science. Likewise, one can recognize the fundamental importance of hermeneutics in anthropology and the social sciences without this implying "textualism", the reduction of social life to semiotics, and the uncritical embrace of postmodernism.

What is needed then is an approach that combines science (naturalism) and hermeneutics (humanism)—thus avoiding the extremes of *positivism*, which repudiates hermeneutics and tends towards a reductive materialism, and *textualism*, which repudiates empirical social science and tends towards cultural idealism, some form of social constructionism, or even—the latest avatar of postmodernism—the embrace of animism (neo-paganism) as an ontology. (No less!)

Hermeneutics (or phenomenology) as interpretive understanding has always been intrinsic to anthropology, especially to the fieldwork experience. Indeed, as Bleicher writes, the social sciences contain a "hermeneutic dimension which is both eradicable and fundamental" (1982: 2). For "meaning" is a central category in the study of social and cultural phenomena. A purely positivistic approach to social life, one that tends to completely repudiate hermeneutics (Abel 1953; Skinner 1953) is therefore untenable. Critical realists, in particular, have always stressed the importance of combining hermeneutics with causal (historical) explanations. Engaging in hermeneutics—understanding the meaning and relevance of other people's words and actions—has thus been described as the "first step" or the "ethnographic moment" for any interpretive understanding or explanation in the social sciences. But while interpretive understanding is both important and necessary, this does not imply that there is no scope for historical analysis and causal explanations (Sayer 2000: 6; Manicas 2006: 65-66).

The interpretation of cultural meanings, social actions, and causal (morphogenetic) explanations of socio-cultural phenomena are therefore complementary and interrelated perspectives, both central to anthropological knowledge. Indeed, as I have stressed elsewhere (Morris 1997: 333-335, 2014A: 406-446), the three key scholars who are often invoked to support the alleged "interpretive turn" in anthropology and the social sciences—Wilhelm Dilthey, Hans-Gorg Gadamer, and Paul Ricoeur—never disavowed the importance of empirical science. To the contrary, they acknowledged the importance of both hermeneutics and empirical science. As Outhwaite suggested with regard to Dilthey's *Lebensphilosophie*, the German scholar "did not believe that the use of verstehen (interpretive understanding) ruled out causal explanation based on comparison and generalizations; the two methods are complementary" (1975: 29).

I have discussed elsewhere at some length the philosophy of positivism, specifically as it related to the sociology of Auguste Comte (2014A: 196-201), and in chapter eight I addressed the anti-realism that was expressed by many logical positivists. Positivism has indeed been the subject of numerous critiques, both within the philosophy of science and within the social sciences. Such critiques long predate the so-called literary or interpretive turn in anthropology. Positivism, however, has a number of features that are worth highlighting again in the present context. The first is that it adopts an empiricist ontology, and thus rejects the distinction between phenomena (the world as it appears to humans) and reality; it thus denies the existence of such indiscernibles as cognitive states, energy, generative processes (or becomings), or causal mechanisms (Bunge 2006: 37). Secondly, it adopts the human conception of causality as involving a regular succession of events, along with the "deductive-normological", or "covering law" model of scientific explanation. Thirdly, in emphasizing the unity of the sciences, positivism tends to give epistemic privilege to mathematics and physics. This often entails a form of reductive naturalism (or mechanistic materialism) (Hempel 1966; Outhwaite 1987: 5-11; Hollis 1994: 40-65).

All these three features of positivism—its phenomenalism, its notion of explanation in terms of empirical generalizations that express universal laws, and its reductive, mechanistic tendencies (physicalism)—have long been critiqued by philosophers of science like Mario Bunge (1996, 2000: 59-63) and by critical realists (Bhaskar 1975; Sayer 2000; Manicas 2006).

What they have suggested as an alternative to the positivist conception of science is a form of understanding or explanation that is *retrodictive*, and thus is focussed not on prediction and control, but rather on identifying the underlying, and often hidden, causal or generative mechanisms (becomings) that explain social facts and cultural phenomena. As Bunge expressed it, to explain a social fact is to exhibit its underlying mechanism; and to show "how it came to be" (1996: 137). Generative mechanisms in the social sciences essentially refer to social processes and are thus fundamentally *historical*. Social and cultural phenomena are the emergent outcomes of real processes which occur in concrete (material) systems—specifically in various kinds of social systems or movements. Many structures, or mechanisms may be involved in such systems, or social forms, and such generative mechanisms imply both human agency, and multiple causal factors—ecological, biological, psychological, social, and cultural. Such generative mechanisms within the social sciences are always processual and historically (and ecologically) situated. Any specific social fact or event can therefore only be understood in terms of a multiplicity of factors (Bunge 2003: 170; Parker 2003: 184-185; Manicas 2006: 4).

All explanations of human social life—whether of religious phenomena, rebellions against colonial rule, or the relation of people to aspects of the natural world—always, therefore, entail a multiplicity of perspectives, as I have illustrated and long contended (Morris 1987, 1998, 2004A, 2017).

One can, of course, acknowledge the importance of scientific knowledge as a creative and imaginative representation of reality, one whose accuracy and validity is tested by various—and different—practices, without accepting either the positivistic conception of science or its deification. Neither Aristotle nor Kant accepted that science was the only form of knowledge, and a realist conception of science has long been articulated. Acknowledging that science is a social practice does not deny the validity of empirical science; equally, the fact that scientific theories and hypotheses are always "mediated" by social practices does not entail a repudiation of the correspondence theory of truth (Reyna 1994).

Literary critics and postmodernists have long been telling us, in oracular fashion, that there is no such thing as the "scientific method", or even that there is no such thing as "science". This is wishful thinking? Nobody has yet suggested that there is no such thing as "poetry" or "literature"! Science, of course, is just a way of understanding the world, and everybody in their practical activities implicitly affirms science.

The scientific method allows us to obtain genuine knowledge about the world in which we live, both natural and social, and is usually described as involving several stages or steps. The first step involves, as Aristotle implied, becoming familiar with all existing or background knowledge in relation to a particular project or subject matter. The second step entails the systematic observation of the phenomena being studied, or the gathering of relevant facts (as data), or even controlled experiments—depending on what aspect of the material world is being studied. Finally, there is the checking or "testing" of a particular representation (or hypothesis or scientific "model"), which has been obtained through empirical procedures, against the real world (or at least aspects of it). Scientific theories, as representations, are never absolute. They are always approximate and provisional, thus open to scrutiny and critique. All science is a collective enterprise, but the validity of scientific knowledge (contra Kuhn 1962 and Rorty 1980) is *not* based on consensus, but on the nature of the world itself—whether natural or social.

Formal science is not simply the extension of common sense, but an effort to take us beyond common-sense understanding (hermeneutics, phenomenology, naïve realism). Using disciplined imagination, it seeks to *explain* phenomena, or even seeks to explore realities (like chromosomes or bacteria) that are not a constituent party of everyday experience. Science, I think, cannot be *equated* with power, the male gender, western culture, or colonialism. It is, however, worth noting the "ethos" of science, as famously outlined by Robert Merton (1973), includes the values of universalism, intellectual honesty and integrity, organized scepticism, communism (an emphasis on the sharing of knowledge), and disinterestedness (in the sense of being free of ideological motivation).

In its most basic sense, as I discussed with reference to the Enlightenment, science involves a ratio-empiricist epistemology, combining empirical observations with causal inferences (reason). Science is simply a method for obtaining valid knowledge about *this* world, and as empirical knowledge, or what Paul Richards describes as "people's science" (1985: 155), it constitutes a part of all human societies.

The importance and relevance of science was perhaps best expressed by Colin Tudge when he wrote: "Science is one of the greatest creations of humans. It does not provide us with absolute truths, and certainly offers no royal road to omniscience, but it does provide us with wondrous and unending insights into the workings of the material universe." (2003: 185).

All this seems to have been lost among temporary postmodern ethnographers, especially those enchanted by tribal animism, who invariably continue to conflate the material world we inhabit with our conceptions of it (Bunge 1998: 233; Capra and Luisi 2014: 2; for a readable and engaging discussion of science see Dunbar 1995; on the exoticism of the Neo-Animist ethnographers see Nugent 2016).

Anthropology has always tried to maintain a bridge between the natural sciences (especially biology and ecology), and the humanities (history, philosophy, and literature in particular). Long before postmodernism and its poetics, there have been those who denied that the discipline is, or ever could be, a science—this is the contention of Husserl and the Neo-Kantian philosophers. Equally, there have been those, like Radcliffe-Brown and Marvin Harris, who have been adamant that anthropology should be modelled on the natural sciences, often conceived in rather positivistic fashion. They have thus been concerned either with using comparative studies to establish causal laws and make inductive generalizations, or, as the structuralists, with delineating universal structures inherent in the human mind. Often such scholars are anti-history, and downplay human subjectivity and agency.

Many anthropologists, however—especially those with a sense of history—have long tended to occupy the middle ground (Wolf 1964: 88; Bloch 1998: 40). Following a long tradition well expressed by Marx, Durkheim, and Max Weber, they have been concerned with both interpretive understanding and the scientific explanation of social and cultural phenomena, without collapsing either into crude positivism or cultural idealism. They have tried in various ways to unite the Enlightenment with various insights drawn from the romantic tradition.

In a critique of the over-emphasis on cultural interpretations, Tim O'Meara (1989) argued that the explanation of human social life and behaviour has always been an intrinsic part of anthropology (and everyday life!) and that anthropology is therefore a humanistic science. The notion that anthropology is simply a rebellion against the Enlightenment, as suggested by the cultural psychologist Richard Shweder (1984) is quite misleading (c.f. Morris 1986).

Human life is inherently social and meaningful—for there *are* signs of meaning in the universe (Hoffmeyer 1996)—as well as being "enmeshed" or "rooted" in the natural world. Indeed, humans—given the paradox of human life—are as much ecological or earthly beings as they are cultural beings—*Homo Symbolicum* (Cassirer). An understanding of the human condition therefore entails both hermeneutic understanding and interpretation (humanism) as well as explanations in terms of causal mechanisms and historical understanding (naturalism). Anthropological understanding, as I have stressed above, must therefore combine hermeneutics and empirical science, and avoid the one-sided emphasis on either hermeneutics—which, in its extreme form, "textualism" (or ethnographic autobiography), denies any empirical science—or on pure naturalism— which, in its extreme form, "positivism", oblates or downplays cultural meanings and human values. The alternative, however, to positivistic science, or what Bourdieu describes as "objectivism" is not a facile acceptance of neo-romantic hermeneutics that espouses a Neo-Kantian idealist paradigm and radical historicism (cultural relativism), still less the embrace of neo-paganism (animism) as a metaphysic. Anthropology must therefore, I think, continue to follow the tradition of the historical social scientists (Marx, Dilthey, Weber, Boas, Evans-Pritchard) and combine hermeneutics (interpretive *understanding*) with empirical science (historical *explanations*). As the radical empiricist Michael Jackson writes: "People cannot be reduced to texts any more than they can be reduced to objects" (1989: 184; for reaffirmations of scientific anthropology in response to the more extreme version of postmodernist hermeneutics see Lett 1997; Kuznar 1997).

It is worth noting that these scholars—as historical anthropologists—though fundamentally historical thinkers, also acknowledged, if only implicitly, that there was an intrinsic *ecological* context to human life.

But acknowledging a naturalistic perspective, the importance of empirical science, and the psychic, moral, and epistemological unity of mankind (universalism), does *not* imply the "destruction" of the concrete, the specificity and diversity of human cultures, or the historical; nor does it imply that people's behaviour is the same everywhere, as Hollinger seems to believe (1994: 67). Nor does a scientific or universalist approach imply a "decontextualized" perspective (Hornborg 1996: 50-55)—the conflation of Enlightenment universalism with so-called "modernity" is misleading. To the contrary, a (historical) universalist approach situates material things, organisms, people, and social facts in a wider human (social) and earthly (ecological) *context*.

As I have long argued, anthropology, despite its undoubted diversity, has a certain unity of purpose and vision. It is perhaps unique among the human sciences in both putting an emphasis and value on cultural difference, thus offering a cultural critique of Western capitalism and its culture, *and* in emphasizing people's shared humanity. Thus, on the one hand, anthropology and ethnographic studies have greatly expanded and enhanced, our understanding of human societies—the diversity and complexity, for example, of kinship systems, modes of agricultural production, the forms of various civilizations, and the variety and complexity of religious institutions. On the other hand, however, in emphasizing our shared humanity, anthropology has indicated the importance of cultural universals, enhanced our sense of moral community, and situated humans squarely within nature. Anthropology has therefore offered insightful accounts of the origins of the human species, on the evolution of human social life and symbolic culture, and on the essential continuity and dialectical inter-relationship between the natural world and human social life and culture. As an academic discipline, anthropology has always placed itself as a comparative historical social science that is firmly based on ethnographic studies (I.M. Lewis 1999: 116)—as I have earlier stressed—at the "interface" between the natural sciences—especially biology—and the humanities (history, philosophy, literature, and the arts).

Sadly, towards the end of the last century, given the arrogant and intolerant rhetoric of postmodern anthropologists who seemed to repudiate empirical science entirely, and the equally dismissive attitude some positivist anthropologists have towards hermeneutics—bolstered by the rise of Neo-Darwinian theory (Tyler 1986; Gellner 1995)—a "wide chasm" seems to have emerged between the various intellectual traditions (Borofsky 1994: 3). Indeed, Maurice Bloch, while emphasizing the "dual heritage" of anthropology, came to bewail the spirit of "fundamentalism" that had developed in anthropology towards the close of the twentieth century. One type of fundamentalism, associated with hermeneutic and postmodernist scholars, conceives of anthropology as a purely "literary enterprise" and repudiates empirical social science. Meanwhile, other anthropologists, taking their intellectual bearings from sociobiology and cognitive science, are aggressively naturalistic and wish to "purify" anthropology of the other orientation, thus reducing anthropology to a branch of biology (Bloch 1998: 39-41).

Anthropology, however, should not be equated with ethnography nor is the fieldwork experience some kind of initiation into some esoteric cult. It is rather a "fact-finding" mission, providing essential data in the development of anthropology as a historical (humanistic) science, and thus enhancing our understanding of the human condition.

As I indicated in the first chapter, the aim of this manifesto has been to outline and advocate a form of dialectical (historical) naturalism: one that dialectically affirms and unites both sides of the "great divide" in anthropology (Fearn 2008). I have therefore sought to acknowledge and extol both humanism (with its emphasis on history and culture) and naturalism (with its emphasis on science and ecology); both the lineaments of the left side of the brain (with its emphasis on focussed attention, analysis, language, and material entities) and the right side of the brain (with its emphasis on patterns, images, synthesis, representations, and relationships); and, finally, as an epistemology, both hermeneutics and the interpretation of cultural meanings, and empirical science, which, going beyond phenomenology, is concerned with the explanation of social life and culture.

I thus aimed to provide a coherent metaphysics—evolutionary or historical naturalism—to complement and support what Mario Bunge described as "the most basic and comprehensive of all the social sciences". Indeed, "Nothing human is alien to anthropology" (1998: 47).

Bibliography

Abel, T. 1953. The Operation Called Verstehen. In H. Feigl and M. Brodbeck (eds) *Readings in the Philosophy of Science*. New York: Appleton-Century-Crofts. pp 677-687.

Abram, D. 1996. *The Spell of the Sensuous*. New York: Random House.

Abu-Lughod, L. 1991. Writing Against Culture. In R.G. Fox (ed) *Recapturing Anthropology*. Santa Fe: School of American Research. pp 137-162.

Ackrill, J.L. 1981. *Aristotle the Philosopher*. Oxford: Clarendon Press.

-----. (ed) . 1987. *The New Aristotle Reader*. Oxford: Clarendon Press.

Adorno, T. and M. 1973. *Dialectic of Enlightenment*. Horkheimer London: Allen Lane.

Althusser, L. 1977. *For Marx*. (Original 1965). London: New Left Books.

-----. 2013. *The Humanist Controversy and Other Writings*. London: Verso.

Archer, M. 1995. *Realist Social Theory: A Morphogenetic Approach*. Cambridge: Cambridge University Press.

-----. 2000 *Being Human: The Problem of Agency*. Cambridge: Cambridge University Press.

Archer, M. et al (eds). 1998. *Critical Realism: Essential Readings*. London: Routledge.

Arendt, H. 1978. *The Life of Mind*. New York: Harcourt Press.

Aristotle. 1962. *The Politics*. (Trans T. Sinclair). Harmondsworth: Penguin Books.

-----. 1965-1970. *History of Animals*. (Trans A. Peck). Cambridge, Mass.: Harvard University Press.

-----. and Hugh Lawson-Tancred. 1986. *De Anima (On the Soul)*. Introd. H. Lawson-Tancred. London: Penguin Books.

Armstrong, D.M. 1997. *A World of States of Affair*. Cambridge: Cambridge University Press.

Assiter, A. 1996. *Enlightened Woman*. London: Routledge.

Audi, R. 2007. *Moral Value and Human Diversity*. Oxford: Oxford University Press.

Augustine, St. 1991. *Confessions*. (Trans H. Chadwick). Oxford: Oxford University Press.

Ayer, A.J. 1936. *Language, Truth and Logic*. (1971 Ed). Harmondsworth: Penguin Books.

Badcock, C. 2000. *Evolutionary Psychology: A Critical Introduction*. Cambridge: Polity Press.

Badiou, A. 2000. *Deleuze: The Clamour of Being*. Minneapolis: University of Minnesota Press.

-----. 2009. *Pocket Pantheon: Figures in Postwar Philosophy*. London: Verso.

Baert, P. 2005. *Philosophy at the Social Sciences: Towards Pragmatism*. Cambridge: Polity Press.

-----. and F. Carreirara Siha. 2010. *Social Theory in the Twentieth Century and Beyond* (2nd Ed). Cambridge: Polity Press.

Barkow, J., L. Cosmides, and J. Tooby, (eds) . 1992. *The Adapted Mind: Evolutionary Psychology and the Generation of Cosmides Culture*. New York: Oxford University Press.

Barnard, A. 2012. *Genesis of Symbolic Thought*. Cambridge: Cambridge University Press.

-----. 2016. *Language in Prehistory*. Cambridge: Cambridge University Press.

Barnes, B. 1983. On the Conventional Character of Knowledge and Cognition. In K. Knorr-Cetina and M. Mulkay (eds). *Science Observed: Perspectives on the Social Study of Science.* London: Sage. pp 19-51.

Bateson, G. 1972. *Steps to an Ecology of Mind.* Chicago: University of Chicago Press.

-----. 1979 *Mind and Maths: A Necessary Unity.* London: Fontana.

-----. and M.C. Bateson. 1987. *Angel's Fear: An Investigation into the Nature and Meaning of the Sacred.* London: Rider.

Becker, G.S. 1976. *The Economic Approach to Human Behaviour.* Chicago: University of Chicago Press.

Benedict, R. 1934. *Patterns of Culture.* London: Routledge & Kegan Paul.

Benthall, J. 1995. From Self-Applause Through Self-Criticism to Self-Confidence. In A.S. Ahmed and C. Shore (eds) *The Future of Anthropology.* London: Athlone. pp 1-11.

Benton, T. 1993. *Natural Relations.* London: Verso.

Bergson, H. 1907. *Creative Evolution.* (1975 Ed). Westport, Conn.: Greenwood Press.

Berlin, I. (ed). 1956. *The Age of Enlightenment: The Eighteenth Century.* New York: Mentor.

-----. 1979. *Against the Current Essays in the History of Ideas.* Oxford: Clarendon Press.

-----. 1990. *The Crooked Timber of Humanity.* London: Fontana.

Bhaskar, R. 1975. *A Realist Theory of Science.* (1978 Ed). Hassocks: Harvester Press.

-----. 1991. *Philosophy and the Idea of Freedom.* Oxford: Blackwell.

Blackmore, S. 1999. *The Meme Machine.* Oxford: Oxford University Press.

Bleicher, J. 1982. *The Hermeneutic Imagination.* London: Routledge & Kegan Paul.

Bloch, M. 1998. *How We Think They Think: Anthropological Approaches to Cognition, Memory and Literacy.* Boulder: Westview Press.

-----. 2005. *Essays in Cultural Transmission.* Oxford: Berg.

-----. 2012. *Anthropology and the Cognitive Challenge.* Cambridge: Cambridge University Press.

Bloor, D. 1976. *Knowledge and Social Imagery.* London: Routledge & Kegan Paul.

Boas, F. 1940. *Race, Language and Culture.* (1982 Ed). Chicago: University of Chicago Press.

Bobbio, N. 1996. *The Age of Rights.* Cambridge: Polity Press.

Bock, K. 1980. *Human Nature and History: A Response to Sociobiology.* New York: Columbia University Press.

Bogdanor, V. 2003. Tract of a Cult Leader Whose Disciples Lost Their Religion. *Times Higher Education.* November 28th.

Bohr, N. 1958. *Atomic Physics and Human Knowledge.* New York: Wiley.

Boisvert, R.D. 1998. *John Dewey: Rethinking Our Time.* Albany: State University of New York Press.

Bookchin, M. 1971. *Post-Scarcity Anarchism.* London: Wildwood House

-----. 1980. *Toward an Ecological Society.* Montreal: Black Rose Books

-----. 1982. *The Ecology of Freedom.* (1991 Rev. Ed). Montreal: Black Rose Books.

-----. 1986. *The Modern Crisis.* Philadelphia: New Society Publishers.

-----. 1989. *Re-Making Society.* Boston: South End Press.

-----. 1995A. *The Philosophy of Social Ecology.* (Original 1990). Montreal: Black Rose Books.

-----. 1995B. *Re-Enchanting Humanity*. London: Cassell.

-----. 1999. *Anarchism, Marxism and the Future of the Left*. Edinburgh: AK Press.

Borofsky, R. (ed) 1994. *Assessing Cultural Anthropology*. New York: McGraw-Hill.

Bourdieu, P. 1990. *The Logic of Practice*. (Original 1980). Cambridge: Polity Press.

-----. 1994. *In Other Words: Essays Towards a Reflexive Sociology*. (Original 1987). Cambridge: Polity Press.

-----. 2004. *Science of Science and Reflexivity*. (Original 2001). Cambridge: Polity Press.

-----. and L.J. Wacquant. 1992. *An Invitation to Reflexive Sociology*. Cambridge: Polity Press.

Boyd, R. and P.J. Richardson. 1985. *Culture and the Evolutionary Process*. Chicago: University of Chicago Press.

Braidotti, R. 2013. *The Posthuman*. Cambridge: Polity Press.

Brassier, R. 2007. *Nihil Unbound: Enlightenment and Extinction*. New York: Palgrave MacMillan.

Bronner, S.E. 2004. *Reclaiming the Enlightenment*. New York: Columbia University Press.

Buchler, J. (ed). 1955. *Philosophical Writings of Peirce*. New York: Dover.

Bufe, C. and M. Vertex. (eds.) 2005. *Dreams of Freedom: A Ricardo Flores Magon Reader*. Edinburgh: AK Press.

Buller, D. T. 2006. Evolutionary Psychology: A Critique. In E. Sober (ed) *Conceptual Issues in Evolutionary Biology*. Cambridge, Mass.: MIT Press.

Bunge, M. 1967. *Scientific Research: The Search for Truth, Vol 2*. New York: Springer.

-----. 1996. *Finding Philosophy in Social Science*. New Haven: Yale University Press.

-----. 1998. *Social Science Under Debate: A Philosophical Perspective*. Toronto: University of Toronto Press.

-----. 1999. A *Dictionary of Philosophy*. Amherst: Prometheus Books.

-----. 1999B. *The Sociology-Philosophy Connection*. New Brunswick: Transaction.

-----. 2001A. *Philosophy in Crisis: The Need for Reconstruction*. Amherst: Prometheus Books.

-----. and M. Mahner. (eds.). 2001B. *Scientific Realism: Selected Essays*. ed. M. Mahner. Amherst: Prometheus Books.

-----. 2003. *Emergence and Convergence: Qualitative Novelty, and the Unity of Knowledge*. Toronto: University of Toronto Press.

-----. 2006. *Chasing Reality: Strife Over Realism*. Toronto: University of Toronto Press.

-----. 2010. *Matter and Mind: A Philosophical Enquiry*. New York: Springer.

-----. 2012. *Evaluating Philosophies*. New York: Springer.

Barr, V. 1995. *Social Constructionism*. (2nd Ed 2003). Hove: Routledge.

Buss, D.M. 1994. *The Evolution of Desire: Strategies in Human Mating*. New York: Harper Collins.

-----. 1999. *Evolutionary Psychology: The New Science of Mind*. Boston: Allyn & Bacon.

Callinicos, A. 1997. Postmodernism: A Critical Diagnosis in *Great Ideas Today*. Chicago: Encyclopaedia Brittanica. pp 206-255.

-----. 2006. *The Resources of Critique*. Cambridge: Polity Press.

Capra, F. 1975. *The Tao of Physics*. London: Fontana/Collins.

-----. 1982. *The Turning Point*. London: Fontana.

-----. 1997. *The Web of Life: A New Synthesis of Mind and Matter.* London: Harper Collins.

-----. and P.L. Luisi 2014. *The Systems View of Life: A Unifying Vision.* Cambridge: Cambridge University Press.

Carnap, R. 1967. *The Logical Structure of the World and Pseudoproblems in Philosophy.* London: Continuum.

Carrithers, M. 1992. *Why Humans Have Culture.* Oxford: Oxford University Press.

Cassirer, E. 1944. *An Essay On Man.* New Haven: Yale University Press.

-----. 1951. *The Philosophy of the Enlightenment.* Princeton, N.J.: Princetown University Press.

Castelnuovo, A. and Kotik-Friedgut, B.S. 2015. *Vygotsky and Bernstein in the Light of the Jewish Tradition.* Boston: Academic Studies Press.

Charlton, N. 2008. *Understanding Gregory Bateson.* Albany: State University of New York Press.

Chattopadhyaya, D. 1959. *Lokayata: A Study in Ancient Indian Materialism.* New Delhi: People's Publishing House.

Chomsky, N. 1975. *Reflections on Language.* New York: Random House.

Chopra, D. 2006. *Life After Death.* London: Rider.

Churchland, P.M. 1984. *Matter and Consciousness.* Cambridge, Mass.: MIT Press.

Cicero, M.T. 1972. *The Nature of the Gods.* London: Penguin Books.

Clark, J.P. and C. Martin. 2004. *Anarchy, Geography, Modernity: The Radical Social Thought and Eliseé Reclus.* Lanham: Lexington Books.

Clements, F. 1949. *The Dynamics of Vegetation.* New York: Wilson.

Clifford, J. and G.E. 1986. *Writing Culture: The Poetics and Politics of Ethnography.* Marcus (eds) Berkeley: University of California Press.

Cohen, P.S. 1968. *Modern Social Theory.* London: Heinemann.

Coleman, J. 1990. *Foundations of Social Theory.* Cambridge, Mass.: Harvard University Press.

Collier, A. 1985. Truth and Practice. In R. Edgeley and R. Osborne (eds). *Radical Philosophy Reader.* London: Verso pp 193-213.

-----. 1994. *Critical Realism: An Introduction to Roy Bhaskar's Philosophy.* London: Verso.

Cosmides, L. and J. Tooby. 1987. From Evolution to Behaviour: Evolutionary Psychology on the Missing Link. In J. Dupre (ed). *The Latest on the Best: Essays and Evolution and Optimality.* Cambridge, Mass.: MIT Press.

Coward, R. and J. Ellis. 1977. *Language and Materialism.* London: Routledge & Kegan Paul.

Coyne, J.A. 2009. *Why Evolution is True.* Oxford: Oxford University Press.

D'Abro, A. 1939. *The Decline of Mechanism in Modern Science.* New York: Van Nostrand.

Dallmayr, T. 1993. *The Other Heidegger.* Ithaca: Cornell University Press.

Daly, M. and M. Wilson. 1998. *Homicide.* New York: Aldine.

Damasio, A. 2000. *The Feeling of What Happens.* London: Vintage.

Darwin, C. 1951. *The Origin of Species.* (Original 1859). London: Oxford University Press.

Davidson, D. 1984. *Inquiries into Truth and Interpretation.* Oxford: Clarendon Press.

Dawe, A. 1979. Theories of Social Action. In T. Bottomore and R. Nisbet (eds). *A History of Sociological Analysis*. London: Heinemann.

Dawkins, R. 1976. *The Selfish Gene*. (1989 Ed). Oxford: Oxford University Press.

-----. 1982. *The Extended Phenotype*. Oxford: Oxford University Press.

-----. 2003. *The Devil's Chaplain: Selected Essays*. London: Weidenfeld & Nicolson.

-----. 2006. *The God Delusion*. London: Bantam Press.

DeLanda, M. 2006. *A New Philosophy of Society: Assemblage Theory and Social Complexity*. London: Continuum.

Deleuze, G. 1990. *Expressionism in Philosophy: Spinoza*. (Original 1968). New York: Zore Books.

-----. 1994. *Difference and Repetition*. (Original 1968). London: Athlone Press.

-----. 1995. *Negotiations 1972-1990*. (Original 1990). New York: Columbia University Press.

-----. 2001. *Pure Immanence: Essays on a Life*. (Original 1995). New York: Zore Books.

-----. 2004. *The Logic of Sense*. (Original 1969). London: Continuum.

-----. and F. Guattari 1977. *Anti-Oedipus: Capitalism and Schizophrenia*. (Original 1972). New York: Viking Press.

----- and F. Guattari. 1988. *A Thousand Plateaus: Capitalism and Schizophrenia*. (Original 1980). London: Athlone Press.

-----. and F. Guattari. 1994. *What is Philosophy?*. (Original 1991). London: Verso.

-----. and C. Parnett. 1987. *Dialogues II*. (Original 1977). London: Continuum.

Dennett, D. 1991. *Consciousness Explained*. Boston: Little, Brown.

-----. 1995. *Darwin's Dangerous Idea: Evolution and the Meaning of Life*. London: Penguin Books.

Derrida, J. 1976. *Of Grammatology*. Baltimore: John Hopkins University.

-----. 1978. *Writing and Difference*. London: Routledge & Kegan Paul.

Desacola, P. 1996. Constituting Nature: Symbolic Ecology and Social Practice. In P. Descola and G. Palsson (eds). *Nature and Society: Anthropological Perspectives*. London: Routledge. pp 82-102.

-----. and G. Palsson. (eds). 1996. *Nature and Society: Anthropological Perspectives*. London: Routledge.

Detmer, D. 2003. *Challenging Postmodernism Philosophy and the Politics of Truth*. Amherst: Humanity Books.

Devall, B. and G. Layton. 1985. *Deep Ecology* Sessions : Peregrine Smith.

Devitt, M. 1984. *Realism and Truth*. (2nd Ed 1991). Oxford: Basil Blackwell.

Dewey, J. 1916A. *Democracy and Education*. New York: MacMillan.

-----. 1916B. *Essays in Experimental Logic*. (1954 Ed). New York: Dover.

-----. 1925. *Experience and Nature*. (1958 Ed). New York: Dover.

-----. 1929. *The Quest for Certainty*. Carbondale: South Illinois Press.

-----. 1934. *Art and Experience*. New York: Minton.

Diamond, J. 1991. *The Rise and Fall of the Third Chimpanzee*. London: Vintage.

Distin, K. 2004. *The Selfish Meme: A Critical Re-Assessment*. Cambridge: Cambridge University Press.

Dodds, E.R. 1977. Plato and the Irrational Soul. In G. Vlastos (ed). *Plato II*. Garden City: Doubleday pp 206-229.

Douglas, M. 1975. *Implicit Meanings*. London: Routledge & Kegan Paul.

Douzinas, C. 2010. Adikia: On Communism and Rights. In C. Douzinas and S. Zizek (eds). *The Idea of Communism*. London: Verso pp 81-100.

Dubos, R. 1973. *A God Within*. London: Sphere Books.

Dumont, L. 1986. *Essays in Individualism*. Chicago: University of Chicago Press.

Dunbar, R. 1995. *The Trouble with Science*. London: Faber & Faber.

-----. 2004. *The Human Story*. London: Faber & Faber.

Dunbar, R., C. Knight, and C. Power (eds). 1999. *The Evolution of Culture*. Edinburgh: Edinburgh University Press.

Dunbar, R, L. Barrett, and J. Lycett. 2007. *Evolutionary Psychology: A Beginner's Guide*. Oxford: Oneworld.

Durham, W. 1979. Toward a Co-Evolutionary Theory of Human Biology and Culture. In N. Chagnon and W. Irons (eds). *Evolutionary Biology and Human Social Behaviour*. North Scituate, Ma.: Duxbury Press. pp 39-58.

-----. 1991. *Co-Evolution: Genes, Culture and Human Diversity*. Stanford: Stanford University Press.

Durkheim, E. 1895. *The Rules of the Sociological Method*. (1982 Ed). London: MacMillan.

-----. 1912. *The Elementary Forms of the Religious Life*. (1984 Ed). London: Allen & Unwin.

Dwyer, P.D. 1996. The Invention of Nature. In R. Ellen and K. Fukui (eds). in *Re-Defining Nature: Ecology, Culture and Domestication*. Oxford: Berg. pp 157-186.

Ehrenfeld, D. 1978. *The Arrogance of Humanism*. Oxford: Oxford University Press.

Ekins, P. 1992. *A New World Order*. London: Routledge.

Eldredge, N. and S.J. Gould. 1972. Punctuated Equilibrium: An Alternative to Phyletic Gradualism. In T.J. Schopf (ed). *Models in Palaeobiology*. San Francisco: Freeman pp 82-115.

Elias, N. 1978. *What is Sociology*. London: Hutchinson.

Elsdon-Baker, F. 2009. *The Selfish Genius: How Richard Dawkins Re-Wrote Darwin's Legacy*. London: Icon Books.

Engels, F. 1940. *Dialectics of Nature*. London: Lawrence & Wishart.

-----. 1969. *Anti-Duhring: Herr Eugen Duhring's Revolution in Science*. (Original 1878). London: Lawrence & Wishart.

Eriksen, T. H, and F.S. 2001. *A History of Anthropology*. Nielsen London: Pluto Press.

Evans-Pritchard, E.E. 1956. *Nuer Religion*. Oxford: Oxford University Press.

-----. 1962. *Essays in Social Anthropology*. London: Faber.

Everett, D. 2012. *Language: The Cultural Tool*. London: Profile Books.

Fabian, J. 1983. *Time and the Other: How Anthropology Makes its Object*. New York: Columbia University Press.

-----. 1994. Ethnographic Objectivity Revisited. In A. Megill (ed). *Rethinking Objectivity*. Durham: Duke University Press.

Fearn, H. 2008. The Great Divide. *Times Higher Education*. 30 November. pp 36-38.

Feigl, H. 1953. The Scientific Outlook: Naturalism and Humanism In H. Feigl and M. Brodbeck (eds). *Readings in the Philosophy of Science*. New York: Appleton-Century. pp 8-18.

Feist, G.J. 2006. *The Psychology of Science and the Origins of the Scientific Mind*. New Haven: Yale University Press.

Ferguson, A. 1995. *An Essay in the History of Civil Society*. (Original 1767). Cambridge: Cambridge University Press.

Fernandez-Armesto, F. 2015. *A Foot in the River: Why our Lives Change – and the Limits of Evolution*. Oxford: Oxford University Press.

Feuerbach, L. 1841. *The Essence of Christianity*. (1957 Ed). New York: Harper & Row.

Feyerabend, P.K. 1975. *Against Method*. London: New Left Books.

Flanagan, O. 1984. *The Science of Mind*. Cambridge, Mass.: MIT Press.

Flax, J. 1990. *Thinking Fragments: Psychoanalysis, Feminism and Postmodernism*. Berkeley: University of California.

Flew, A. 1985. *Thinking about Social Thinking*. London: Fontana.

Fodor, J. 1983. *The Modularity of Mind*. Cambridge, Mass.: MIT Press.

Foltz, B. 1995. *Inhabiting the Earth*. New Jersey: Humanities Press.

Foster, J.B. 2000. *Marx's Ecology: Materialism and Nature*. New York: Monthly Review Press.

-----. 2009. *The Ecological Revolution: Making Peace with the Planet*. New York: Monthly Review Press.

-----. and B. Clark and R. 2008. *Critique of Intelligent Design*. York New York: Monthly Review Press.

-----. 2010. *The Ecological Rift: Capitalism's War on the Earth*. New York: Monthly Review Press.

Foucault, M. 1970. *The Order of Things: The Methodology of the Human Sciences*. London: Tavistock.

-----. 1972. *The Archaeology of Knowledge*. London: Tavistock.

-----. 1980. *Power/Knowledge: Selected Interviews and other Writings 1972-1977*. New York: Pantheon Books.

Freire, P. 1974. *Education: The Practice of Freedom*. London: Writers and Readers Publishing.

French, S. 2014. *The Structure of the World: Metaphysics and Representation*. Oxford: Oxford University Press.

Fromm, E. 1949. *Man for Himself*. London: Routledge & Kegan Paul.

-----. 1955. *The Sane Society*. (1963 Ed). London: Routledge & Kegan Paul.

-----. 1964. *The Heart of Man*. New York: Harper & Row.

Fukuyama, F. 1992. *The End of History and the Last Man*. London: Hamish Hamilton.

Gadamer, H-G. 1976. *Philosophical Hermeneutics*. Berkeley: University of California Press.

Gare, N. 2004. *The Future of Social Theory*. London: Hutchinson.

Garfinkel, H. 1967. *Studies in Ethnomethodology*. Englewood Cliffs: Prentice Hall.

Gay, P. 1966. *The Enlightenment: An Interpretation: The Rise of Modern Paganism*. New York: Norton.

-----. 1969. *The Enlightenment: An Interpretation: The Science of Freedom*. New York: Norton.

Gee, H. 2000. *Keep Time: Cladistics, the Revolution in Evolution*. London: Fourth Estate.

Geertz, C. 1973. *The Interpretation of Culture*. New York: Basic Books.

Gellner, E. 1985. *Relativism and the Social Sciences*. Cambridge: Cambridge University Press.

-----. 1992. *Postmodernism, Reason and Religion*. London: Routledge.

-----. 1995. *Anthropology and Politics: Revolutions in the Sacred Grove*. Oxford: Blackwell.

Gergen, K.J. 1991. *The Saturated Self: Dilemmas of Identity in Contemporary Life*. New York: Basic Books.

Gerth, H. and C.W. Mills (ed). 1948. *From Max Weber: Essays in Sociology*. London: Routledge & Kegan Paul.

Gibson, J.J. 1979. *The Ecological Approach to Visual Perception*. Boston: Houghton Mifflin.

Gibson, T. and K. Sillander (eds). 2011. *Anarchic Solidarity: Autonomy, Equality and Fellowship in Southeast Asia*. New Haven: Yale University Press.

Giddens, A. 1984. *The Constitution of Society*. Cambridge: Polity Press.

-----. 1996. *In Defence of Sociology.* Cambridge: Polity Press.

Giere, R.N. 2006. *Scientific Perspectivism*. Chicago: University of Chicago Press.

Gleick, J. 1988. *Chaos: Making a New Science*. London: Vintage.

Goldman, L. 1977. *Cultural Creation in Modern Society*. Oxford: Blackwell.

Goodman, N. 1978. *Ways of Worldmaking*. Indianapolis: Hackett.

Goodwin, B. 1994. *How the Leopard Changed its Spots: The Evolution of Complexity*. London: Orion Books.

Gotthelf, A. and J.G. Lennox (eds). 1987. *Philosophical Issues in Aristotle's Biology*. Cambridge: Cambridge University Press.

Gould, S.J. 1980. *Ever Since Darwin: Reflections on Natural History*. London: Penguin Books.

-----. 1983. *The Panda's Thumb*. London: Penguin Books.

-----. 1984. *Hen's Teeth and Horses Toes*. London: Penguin Books.

-----. 1987. *Time's Arrow, Time's Cycle: Myth and Metaphor in the Discovery of Geological Time*. London: Penguin Books.

-----. 1989. *Wonderful Life: Burgess Shale and the Nature of History*. London: Penguin Books.

-----. 2003. *The Hedgehog, The Fox and The Magister's Pox*. London: Cape.

-----. P. McGarr and S. Rose (eds). 2006. *The Richness of Life: The Essential Stephen Jay Gould*. London: Cape

Graham, J. 1990. Theory and Essentialism in Marxist Geography. *Antipode* 22: 53-66.

Granovetter, M. 1985. Economic Action and Social Structure: The Problem of Embeddedness. *American Journal of Sociology* 91: 481-510.

Gratton, P. 2014. *Speculative Realism: Problems and Prospects*. London: Bloomsbury.

Gray, J. 2002. *Straw Dogs: Thoughts on Humans and other Animals*. London: Granta Books.

Grayling, A.C. 2001. *Wittgenstein: A Very Short Introduction*. Oxford: Oxford University Press.

-----. 2007. *Truth, Meaning and Realism*. London: Continuum.

-----. 2009. *Ideas that Matter: Personal Guide for the 21st Century*. London: Weidenfeld & Nicolson.

Greenwood, J.D. 1991. *Relations and Representations*. London: Routledge.

Griffiths, P. and R.D. Gray. 1994. Developmental System and Evolutionary Explanations. *Journal of Philosophy* 9: 277-304.

Gross, E. 2008. *Chaos, Territory, Art: Deleuze and the Framing of the Earth*. New York: Columbia University Press.

Gross, P.R. and N. Levitt. 1998. *Higher Superstition: The Academic Left and its Quarrels and Science*. Baltimore: John Hopkins University Press.

Gruber, H.E. 1974. *Darwin on Man: A Psychological Study of Scientific Creativity*. Chicago: University of Chicago Press.

Guthrie, W.K.C. 1950. *The Greek Philosophers from Thales to Aristotle*. London: Methuen.

Habermas, J. 1987. *The Philosophical Discourse of Modernity*. Cambridge: Polity Press

Hacking, I. 1999. *The Social Construction of What?* Cambridge: Cambridge University Press.

-----. 2002. *Historical Ontology*. Cambridge, Mass.: Harvard University Press.

Hallowell, A.I. 1974. *Culture and Experience*. (Original 1955). Philadelphia: University of Pennsylvania Press.

Hallward, P. 2006. *Out of this World: Deleuze and the Philosophy of Creation*. London: Verso.

Hampson, N. 1968. *The Enlightenment*. London: Penguin Books.

Hann, C. and K. Hart. 2011. *Economic Anthropology: History, Ethnography, Critique*. Cambridge: Polity Press.

Hann, S. 2010. Postmodern Social Theory. In A. Elliott (ed). *The Routledge Companion to Social Theory*. London: Routledge pp 117-134.

Harman, G. 2010. *Towards Speculative Realism: Essays and Lectures*. Winchester: Zone Books.

Harre, R. 1970. *Principles of Scientific Thinking*. Chicago: University of Chicago Press.

-----. 1983. *Personal Being: A Theory of Individual Psychology*. Oxford: Blackwell.

Harries-Jones, P. 1995. *A Recursive Vision: Ecological Understanding and Gregory Bateson*. Toronto: University of Toronto Press.

Harris, M. 1980. *Cultural Materialism: The Struggle for a Science of Culture*. New York: Columbia University Press.

-----. 1995. Anthropology and Postmodernism. In M.F. Murphy and M. Margolis (eds). *Science, Materialism, and the Study of Culture*. Gainesville, Fla: University Press of Florida. pp 62-77.

Harvey, D. 2016. *The Ways of the World*. London: Profile Books.

Hasan, R. 2010. *Multiculturalism: Some Inconvenient Truths*. London: Methuen.

Hastrup, K. 1995. *A Passage to Anthropology: Between Experience and Theory*. London: Routledge.

Hebb, D.O. 1980. *Essay on Mind*. Hillsdale, N.J.: Erlbaum.

Heidegger, M. 1962. *Being and Time*. (Original 1927). Oxford: Blackwell.

Heidegger, M. 1975. *Poetry, Language, Thought*. New York: Harper & Row.

-----. 1978. *Basic Writings*. ed. D.F. Krell. (1993 Ed). London: Routledge.

-----. 1988. *The Basic Problems of Phenomenology*. Bloomington: Indiana University Press.

-----. 1994. *Basic Questions of Philosophy*. Bloomington: Indian University Press.

Heisenberg, W. 1958. *Physics and Philosophy*. New York: Harper.

Hempel, C.G. 1966. *Philosophy of Natural Science*. Englewood Cliffs, N.J.: Prentice Hall.

Himmelfarb, G. 2008. *The Road to Modernity*. London: Vintage Books.

Hindess, B. and P. Hirst 1977. *Modes of Production and Social Formation*. London: MacMillan.

Hirschfeld, L.A. and S.A Gelman (eds). 1994. *Mapping the Mind: Domain Specificity in Cognition and Culture*. Cambridge: Cambridge University Press.

Hoffmeyer, J. 1996. *Signs of Meaning in the Universe*. Bloomington: Indiana University Press.

Holbraad, M. 2012. *Truth in Motion: The Recursive Anthropology of Cuban Divination*. Chicago: University of Chicago Press.

Hollinger, R. 1986. Ayn Rand's Epistemology in Historical Perspective. In D.J. Den Uyl and D.B. Rasmussen (eds). *The Philosophical thought of Ayn Rand*. Urbana: University of Chicago Press. pp 38-59.

-----. 1994. *Postmodernism and the Social Sciences: A Thematic Approach*. London: Sage.

Hollis, M. 1994. *The Philosophy of Social Science: An Introduction*. Cambridge: Cambridge University Press.

Homans, G.C. 1974. *Social Behaviour: Its Elementary Forms*. New York: Harcourt Brace.

Hook, S. 1971. *From Hegel to Marx*. (Original 1962). Ann Arbor: University of Michigan Press.

Hornborg, A 1996. Ecology as Semiotics: Outlines of a Contextualist Paradigm for Human Ecology. In P. Descola and G. Pálsson (eds). *Nature and Society: Anthropological Perspectives*. Routledge. pp 45-62.

Horowitz, I. 1983. *C. Wright Mills: An American Utopian*. New York: Free Press.

Hull, D.L. 1978. The Matter of Individuality. *Philosophy of Science* 45/3: 335-360.

Hume, D. 1985. *Essays: Moral, Political and Literary*. (Original 1777). Indianapolis: Liberty Fund.

Hunter, M. 1936. *Reaction to Conquest*. (1961 Ed). Oxford: Oxford University Press.

Husserl, E. 1931. *Cartesian Meditations*. (1993 Ed). Dordrecht: Kluwer.

-----. 1970. *The Crisis of European Sciences and Transcendental Phenomenology*. Evanston: Northwestern University Press.

Hyland, P. (ed). 2003. *The Enlightenment: A Sourcebook and Reader*. London: Routledge.

Ingold, T. (ed). 1990. *The Concept of Society is Theoretically Obsolete*. University of Manchester: Group for Debates in Anthropological Theory.

-----. 2000. The Poverty of Selectionism. *Anthropology Today* 16: 1-2.

-----. 2007. *Lines: A Brief History*. London: Routledge.

-----. 2011. *Being Alive: Essays in Movement, Knowledge and Description*. London: Routledge.

-----. 2013A. Prospect. In T. Ingold and G. Palsson (eds). *Biosocial Becomings: Integrating Social and Biological Anthropology*. Cambridge: Cambridge University Press. pp 1-21.

-----. 2013B. Anthropology Beyond Humanity. *Finnish Anthropological Society*: 38/3: 15-23.

Ingold, T. and G. Palsson (eds). 2013. *Biosocial Becomings: Integrating Social and Biological Anthropology*. Cambridge: Cambridge University Press.

Israel, J. 2001. *Radical Enlightenment: Philosophy and the Making of Modernity 1650-1750*. Oxford: Oxford University Press.

-----. 2006. *Enlightenment Contested: Philosophy, Modernity and the Emancipation of Man 1670-1752*. Oxford: Oxford University Press.

Jablonka, E. and M.A. 2014. *Evolution in Four Dimensions: Genetic Epigenetic,* Lamb *Behavioural and Symbolic Variation in the History of Life*. (Original 2005). Cambridge, Mass.: Harvard University Press.

Jackson, M. 1989. *Paths Towards a Clearing*. Bloomington: Indiana University Press.

Jalais, A. 2010. *Forest of Tigers: People, Politics and Environment in the Sundarbans*. Abingdon: Oxford University Press.

Jameson, F. 1998. *The Cultural Turn: Selected Writings on the Postmodern 1983-1998*. London: Verso.

Jenkins, R. 1992. *Pierre Bourdieu*. London: Routledge.

-----. 2002. *Foundations of Sociology*. Basingstoke: Palgrave MacMillan.

-----. 2008. *Social Identity*. (Original 1996). London: Routledge.

Johnson, M. 1987. *The Body in the Mind*. Chicago: University of Chicago Press.

Jonas, H. 1966. *The Phenomenon of Life: Toward a Philosophical Biology*. Chicago: University of Chicago Press.

Jovchelovitch, S. 2007. *Knowledge in Context: Representation, Community and Culture*. London: Routledge.

Kamenka, E. 1970. *The Philosophy of Ludwig Feuerbach*. London: Routledge & Kegan Paul.

Kant, I. 1781. *The Critique of Pure Reason*. (1959 Ed). London: Dent.

-----. 2007. *Anthropology, History and Education*. Cambridge: Cambridge University Press.

Keesing, R. 1994. Theories of Culture Re-Visited. In R. Borofsky (ed). *Assessing Cultural Anthropology*. New York: McGraw-Hill. pp 301-310

Kingdom, J. 1992. *No Such Thing as Society? Individualism and Community*. Buckingham: Open University Press.

Kitcher, P. 1985. *Vaulting Ambitions: Sociobiology and the Quest for Human Nature*. Cambridge, Mass.: MIT Press.

-----. 2009. *Living with Darwin*. Oxford: Oxford University Press.

Knight, C. 2016. *Decoding Chomsky: Science and Revolutionary Politics*. New Haven: Yale University Press.

Knorr-Cetina, K. and M. Mulkay (eds). 1983. *Science Observed: Perspectives in the Social Study of Science*. London: Sage

Koch, A.M. 2011. Post-Structuralism and the Epistemological Basis of Anarchism. In D. Rousselle and S. Evren (eds). *Post-Anarchism: A Reader*. London: Pluto. pp 23-40.

Kohan, N. 2005. Postmodernism, Commodity, Fetishism and Hegemony. *International Socialism*. 105: 139-158.

Kohn, E. 2013. *How Forests Think: Toward an Anthropology Beyond the Human*. Berkeley: University of California Press.

Kolakowski, L. 1978. *Main Currents in Marxism, Vol. 3*. Oxford: Oxford University Press.

Korzybski, A. 1949. *Science and Society*. Lakeville, Conn.: International Library.

Kovel, J. 2002. *The Enemy of Nature*. London: Zed Books.

Kramnick, I. (ed). 1995. *The Portable Enlightenment Reader*. London: Penguin Books.

Kroeber, A. 1917. The Superorganic. *American Anthropologist*. 19: 163-213.

Kuhn, T.S. 1962. *The Structure of Scientific Revolutions*. Chicago: University of Chicago Press.

Kumar, S. 2016. One Earth, One Humanity, One Future. *Resurgence & Ecologist* 294: 44-45

Kuper, A. 1973. *Anthropologists and Anthropology*. London: Penguin Books.

Kuper, A. 1994. *The Chosen Primate: Human Nature and Cultural Diversity*. Cambridge, Mass.: Harvard University Press.

-----. 1999. *Culture: The Anthropologists Account*. Cambridge, Mass.: Harvard University Press.

Kurtz, P. 1983. *In Defence of Secular Humanism*. Amherst: Prometheus Books.

Kurzweil, E. 1980. *The Age of Structuralism: Levi-Strauss to Foucault*. New York: Columbia University Press.

Kuznar, L.A. 1997. *Reclaiming Scientific Anthropology*. Walnut Creek: Atlanta Press.

Lakoff, G. and M. Johnson. 1999. *Philosophy of the Flesh: The Embodied Mind and its Challenge to Western Thought*. New York: Basic Books.

Laland, K.N. and G. Brown. 2002. *Sense and Nonsense: Evolutionary Perspectives in Human Behaviour*. Oxford: Oxford University Press.

Latour, B. and S. Woolgar. 1979. *Laboratory Life: The Construction of Scientific Facts*. (1986 Ed) Princeton: Princeton University Press.

-----. 1987. *Science in Action*. Cambridge, Mass.: Harvard University Press.

-----. 1988. *The Pasteurization of France*. Cambridge, Mass.: Harvard University Press.

-----. 1993. *We Have Never Been Modern*. London: Prentice Hall.

-----. 2004A. *Politics of Nature*. Cambridge, Mass.: Harvard University Press.

-----. 2004B. Why Has Critique Run Out of Steam: From Matters of Fact to Matters of Concern. *Critical Enquiry* 30: 225-248.

-----. 2005. *Reassembling the Social: An Introduction to Actor Network Theory*. Oxford: Oxford University Press.

-----. G. Harman and P. Erdely. 2011. *The Prince and the Wolf: Latour and Harman at the L.S.E.* Winchester: Zero Books.

Layder, D. 1994. *Understanding Social Theory*. London: Sage.

Leahey, T.H. 1987. *A History of Psychology*. Englewood Cliffs: Prentice Hall.

Lehman, D. 1991. *Signs of the Times: Deconstruction and the Fall of Paul de Man*. London: Deutsche.

Lenin, V. 1972. *Materialism and Empirio-Criticism*. (Original 1909). Peking: Foreign Languages Press.

Leroi, A.M. 2014. *The Lagoon: How Aristotle Invented Science*. London: Bloomsbury.

Lett, J. 1997. *Science, Reason and Anthropology*. Lanham: Rowman & Littlefield.

Levins, R. and R.C. 1985. *The Dialectical Biologist*. Lewontin Cambridge, Mass.: Harvard University Press.

Levi-Strauss, C. 1966. *The Savage Mind*. London: Weidenfeld & Nicolson.

-----. 1976. *Tristes Tropiques*. (Original 1955). London: Penguin Books.

Levy, H. 1938. *A Philosophy for a Modern Man*. London: Gollancz.

Lewes, G.H. 1875. *Problems of Life and Mind*. London: Trubner.

Lewis, D. 1986. *On the Plurality of Worlds*. Oxford: Blackwell.

Lewis, H.S. 2014. *In Defence of Anthropology*. New Brunswick: Transaction Publishers.

Lewis, I.M. 1999. *Arguments with Ethnography: Comparative Approaches to History, Politics and Religion*. London: Athlone Press.

Lewis, J. 1962. *History of Philosophy*. London: English Universities Press.

Lewontin, R.C. and R. Levins. 2007. *Biology under the Influence: Dialectical Essays on Biology, Agriculture and Health*. New York: Monthly Review Press.

Lichtheim, G. 1971. *From Marx to Hegel and Other Essays*. New York: Seabury Press.

Lichtheim, G. 1972. *Essays in the Twentieth Century*. (2000 Ed). London: Phoenix Press.

Llobera, J.R. 1994. *The God of Modernity*. Oxford: Berg.

-----. 2003. *The Making of Totalitarian Thought*. Oxford: Berg.

Lloyd, G.E.R. 1968. *Aristotle: The Growth and Structure of His Thought*. Cambridge: Cambridge University Press.

Lopston, P. 2001. *Theories of Human Nature*. Ontario: Broadview Press.

Lovejoy, A.O. 1936. *The Great Chain of Being*. Cambridge, Mass.: Harvard University Press.

Lucretius 1969. *On the Nature of Things*. Indianapolis: Hackett Publishing.

Lukacs, G. 1971. *History and Class Consciousness*. (Original 1923). London: Merlin Press.

Lukes, S. 1968. Methodological Individualism Reconsidered. *British Journal of Sociology* 19: 119-129.

-----. 1973 *Individualism*. Oxford: Blackwell.

Lumsden, C.J. and E.O. Wilson. 1983. *Promethean Fire: Reflections on the Origin of Mind*. Cambridge, Mass.: Harvard University Press.

Lyotard, J-F. 1984. *The Postmodern Condition*. Manchester: Manchester University Press.

McGarr, P. 1994. Engels and Natural Science. *International Socialism Journal*. 65: 143-176.

McGilchrist, I. 2010. *The Master and His Emissary*. New Haven: Yale University Press.

McGrath, A.E. 2001. *Christian Theology: An Introduction*. Oxford: Oxford University Press.

McLellan, D. 1973. *Karl Marx: His Life and Thought*. London: Granada.

McLellan, D. (ed). 2000. *Karl Marx: Selected Writings*. Oxford: Oxford University Press.

MacMurray, J. 1932. *Freedom in the Modern World*. London: Faber & Faber.

-----. 1933. *Interpreting the Universe*. London: Faber & Faber.

-----. 1957. *Self as Agent*. London: Faber & Faber.

MacPherson, C.B. 1962. *The Political Theory of Political Individualism*. Oxford: Oxford University Press.

Magee, B. 1997. *Confessions of a Philosopher*. London: Weidenfeld & Nicolson.

Mahner, M. and M. Bunge. 1997. *Foundation of Biophilosophy*. New York: Springer.

Malik, K. 2000. *Man, Beast and Zombie*. London: Weidenfeld & Nicolson.

Malinowski, B. 1922. *Argonauts of the Western Pacific*. London: Routledge.

Malpas, S. 2003. *Jean-Francois Lyotard*. London: Routledge.

Manicas, P.T. 2006. *A Realist Philosophy of Social Science.* Cambridge: Cambridge University Press.

Marcus, G.E. 1995. The Re-Design of Ethnography After the Critique of Rhetoric. In B. Goodman and W.R. Fisher (eds). *Rethinking Knowledge.* Albany: State University of New York Press.

-----. and M. J. Fischer 1986. *Anthropology as Culture Critique.* Chicago: University of Chicago Press.

Margulis, L. and D. 1997. *Slanted Truths: Essays on Gaia, Symbiosis and Evolution.* Sagan New York: Springer.

-----. 2007. *Dazzle Gradually: Reflections on the Nature of Nature.* White River Junction: Chelsea Green Publishing.

Marks, T. 1998. *Gilles Deleuze: Vitalism and Multiplicity.* London: Pluto Press.

Marx, K. 1957. *Capital.* (Original 1867). London: Dent.

-----. 1975. *Early Writings.* London: Penguin Books.

-----. and F. Engels. 1956. *The Holy Family.* (Original 1845). London: Lawrence & Wishart.

-----. 1965. *The German Ideology.* London: Lawrence & Wishart.

-----. 1968. *Selected Writings.* London: Lawrence & Wishart.

Maximoff, G.P. (ed). 1953. *The Political Philosophy of Bakunin: Scientific Anarchism.* New York: Free Press.

Mayr, E. 1952. *The Growth of Biological Thought.* Cambridge, Mass.: Harvard University Press.

-----. 1988. *Towards a New Philosophy of Biology.* Cambridge, Mass.: Harvard University Press.

-----. 1991. *One Long Argument.* London: Penguin Books.

-----. 1997. *This is Biology: The Science of the Living World.* Cambridge, Mass.: Harvard University Press.

-----. 2002. *What Evolution Is.* London: Weidenfeld & Nicolson.

-----. 2004. *What Makes Biology Unique.* Cambridge: Cambridge University Press.

Meillassoux, Q. 2008. *After Finitude: An Essay on the Necessity of Contingency.* London: Bloomsbury.

Merleau-Ponty, M. 1964. *Sense and Nonsense.* Evanston, Ill: Northwestern University Press.

Merton, R.K. 1973. *The Sociology of Science.* Chicago: University of Chicago Press.

Midgley, M. 1985. *Evolution as a Religion.* London: Methuen.

-----. 2000 Why Memes? In H. and S. Rose. (eds). *Alas, Poor Darwin: Arguments Against Evolutionary Psychology.* London: Vintage Books. pp 67-84.

Mikics, D. 2009. *Who was Jacques Derrida: An Intellectual Biography.* New Haven: Yale University Press.

Mill, J.S. 1843. *System of Logic.* Charlottesville: Lincoln-Rembrandt.

-----. 1987. *The Logic of the Moral Sciences.* (Original 1843). London: Duckworth.

Miller, D. 1983. *The Pocket Popper.* London: Fontana.

Mills, C.W. 1959. *The Sociological Imagination.* Oxford: Oxford University Press.

Milton, K. 1996. *Environmentalism and Culture Theory.* London: Routledge.

Montesquieu, C. de. 1989. *The Spirit of Laws.* (Original 1748). Cambridge: Cambridge University Press.

Moran, D. 2005. *Edmund Husserl: Founder of Phenomenology*. Cambridge: Polity Press.

Morgan, C.L. 1923. *Emergent Evolution*. London: Williams and Norgate.

Morris, B. 1981. Changing Views of Nature. *The Ecologist* 2: 130-137.

-----. 1982. *Forest Traders: A Socio-Economic Study of the Hill Pandaram*. London: Athlone Press.

-----. 1985. The Rise and Fall of the Human Subject. *Man* 20: 722-742.

-----. 1986. Is Anthropology Simply a Romantic Rebellion Against the Enlightenment? *Eastern Anthro*pology 39: 359-364.

-----. 1987. *Anthropological Studies of Religion*. Cambridge: Cambridge University Press.

-----. 1990A. Indian Materialism. *The Secularist* (Pune) 123: 63-72.

-----. 1990B. The Dualistic Worldview of Tim Ingold. *Eastern Anthro*pology 43: 265-274.

-----. 1991. *Western Conceptions of the Individual*. Oxford: Berg.

-----. 1993. Rajneesh: The Failed Guru. *The Sceptic* 7/3: 10-12.

-----. 1994. *Anthropology of the Self: The Individual in Cultural Perspective*. London: Pluto Press.

-----. 1996. *Ecology and Anarchism: Essays and Reviews or Contemporary Thought*. Malvern Wells: Images.

-----. 1997. In Defence of Realism and Truth: Critical Reflections on the Anthropological Followers of Heidegger. *Critique of Anthropology* 17/3: 313-340.

-----. 1998. *The Power of Animals: An Ethnography*. Oxford: Berg.

-----. 2000. *Animals and Ancestors: An Ethnography*. Oxford: Berg.

-----. 2004A. *Insects and Human Life*. Oxford: Berg.

-----. 2004B. *Kropotkin: The Politics of Community*. Amherst: Humanity Books.

-----. 2006. *Religion and Anthropology: A Critical Introduction*. Cambridge: Cambridge University Press.

-----. 2012. *Pioneers of Ecological Humanism*. Brighton: Book Guild.

-----. 2014A. *Anthropology and the Human Subject*. Bloomington: Trafford Publishing.

-----. 2014B. *Anthropology, Ecology and Anarchism: A Reader*. Oakland, Ca.: PM Press.

-----. 2015. Anthropology of Anarchist Solidarity. *Anthropology Today*. 31/3: 19-20.

-----. 2016. *An Environmental History of Southern Malawi: Land and People of the Shire Highlands*. Basingstoke: Palgrave MacMillan.

-----. 2017. *Visions of Freedom: Critical Writings on Anarchism and Ecology*. Montréal: Black Rose Books.

Moser, P.K. and T.D. Trout (eds). 1995. *Contemporary Materialism: A Reader*. London: Routledge.

Mosko, M.S. and F.H. Damon. 2005. *On the Order of Chaos: Social Anthropology and the Science of Chaos*. Oxford: Berghahn.

Mouzelis, N.P. 2008. *Modern and Postmodern Social Theorizing: Bridging the Divide*. Cambridge: Cambridge University Press.

Mumford, L. 1944. *The Condition of Man*. London: Secker & Warburg.

-----. 1951. *The Conduct of Life*. London: Secker & Warburg.

Mumford, S. 2012. *Metaphysics: A Very Short Introduction* Oxford: Oxford University Press.

Munck, T. 2000. *The Enlightenment: A Comparative Social History 1721-1794*. London: Arnold.

Murdoch, I. 1992. *Metaphysics as a Guide to Morals*. London: Penguin Books.

Murphy, M.F. and M.L. Margulis (eds). 1995. *Science, Materialism, and the Study of Culture*. Gainesville: University Press of Florida.

Nanda, M. 2003. *Prophets Facing Backward: Postmodern Critiques of Science and Hindu Nationalism in India*. New Brunswick: Rutgers University Press.

Nencel, L. and P. Pels (eds). 1991. *Constructing Knowledge*. London: Sage.

Nielsen, K. 2001. *Naturalized Religion*. Amherst: Prometheus Books.

Nietzsche, F. 1956. *The Birth of Tragedy and the Genealogy of Morals*. (Original 1872/1887). New York: Doubleday.

-----. 1967. *The Will to Power*. New York: Random House.

Nisbet, R. 1986. *Conservation: Dream and Reality*. Milton Keynes: Open University Press.

Noske, B. 1997. *Beyond Boundaries: Humans and Animals*. Montreal: Black Rose Books.

Novack, G. 1975. *Pragmatism Versus Marxism: An Appraisal of John Dewey's Philosophy*. New York: Pathfinder Press.

Novack, G. 1978. *Polemics in Marxist Philosophy*. New York: Monad Press.

Nugent, S. (ed). 2012. *Critical Anthropology: Foundational Works*. Walnut Creek, Ca.: Life Coast Press.

-----. 2016. The Exotic Albatross: Exotic Indians, Exotic Theory. In B. Kapferer and D. Theodossopoulos (eds). *Beyond Exoticism*. Oxford: Berghahn. pp 65-83.

Okasha, S. 2002. *Philosophy of Science: A Very Short Introduction*. Oxford: Oxford University Press.

O'Keefe, T. 2010. *Epicureanism*. Durham: Acumen.

Okin, S.M. 1979. *Woven in Western Political Thought*. Princeton: Princeton University Press.

O'Meara, T. 1989. Anthropological as Empirical Science. *American Anthropologist* 91: 354-369.

Outhwaite, W. 1975. *Understanding Social Life*. London: Allen & Unwin.

-----. 1987. *New Philosophies of Social Science*. London: MacMillan.

-----. 2006. *The Future of Society*. Oxford: Blackwell.

Outram, D. 1995. *The Enlightenment*. Cambridge: Cambridge University Press.and R.D. Gray (eds)

Oyama, S. 1985. *The Ontogeny of Information*. Cambridge: Cambridge University Press.

Oyama, S, P.E. Griffiths, and R.D. Gray (eds). 2001. *Cycles of Contingency: Development Systems and Evolution*. Cambridge, Mass.: MIT Press.

Padover, S.K. (ed). 1979. *The Letters of Karl Marx*. Englewood Cliffs: Prentice Hall.

Pagden, A. 2013. *The Enlightenment and Why it Matters*. Oxford: Oxford University Press.

Palsson, G. 1996. Human-Environment Relations: Orientalism Paternalism and Communalism. In P. Descola and G. Palsson (eds). *Nature and Society: Anthropological Perspectives*. London and New York: Routledge. pp 63-81.

Parker, J. 2003. *Social Theory: A Basic Tool Kit*. Basingstoke: Palgrave MacMillan.

Parson, T. 1937. *The Structure of Social Action*. New York: McGraw-Hill.

-----. 1951. *The Social System*. New York: Free Press.

Patterson, T.C. 1997. *Inventing Western Civilization*. New York: Monthly Review Press.

-----. 2009. *Karl Marx: Anthropologist*. Oxford: Berg.

Peikoff, L. 1991. *Objectivism: The Philosophy of Ayn Rand*. New York: Penguin Books.

Pettit, P. 2011. The Virtual Reality of Homo Economicus. In D. Steele and F. Guala (eds). *The Philosophy of Social Science Reader*. London: Routledge. pp 248-261.

Pfeiffer, J.E. 1982. *The Creative Explosion: An Inquiry into the Origins of Art and Religion*. New York: Harper & Row.

Pinker, S. 1997. *How the Mind Works*. London: Penguin Books.

-----. 2002. *The Blank Slate: The Modern Denial of Human Nature*. London: Penguin Books.

Plato. 1954. *The Last Days of Socrates*. London: Penguin Books.

-----. 1965. *Timaeus*. London: Penguin Books.

Plotkin, H. 1997. *Evolution in Mind: An Introduction to Evolutionary Psychology*. London: Penguin Books.

Plumwood, V. 1993. *Feminism and the Mastery of Nature*. London: Routledge.

Pocock, D. 1961. *Social Anthropology*. London: Steed & Wood.

Popper, K.R. 1945. *The Open Society and Its Enemies: Hegel and Marx. Vol 2.* London: Routledge.

-----. 1959. *The Logic of Scientific Discovery*. (Original 1934). London: Hutchinson.

-----. 1992. *In Search of a Better World*. London: Routledge.

Porter, R. 2001. *The Enlightenment*. Basingstoke: Palgrave.

-----. 2003. *Flesh in the Age of Reason*. London: Penguin Books.

Prigogine, I. and I. 1984. *Order Out of Chaos: Man's New Dialogue with Nature*. Stengers London: Fontana.

Prindle, D.F. 2009. *Stephen Jay Gould and the Politics of Evolution*. Amherst: Prometheus Books.

Psillos, S. 1999. *Scientific Realism: How Science Tracks Truth*. London: Routledge.

Purchase, G. 1995. *A Thesis on Systems Theory*. Sydney: Unpublished MSS.

-----. 1997. *Anarchism and Ecology*. Montreal: Black Rose Books.

Putnam, H. 1990. *Realism with a Human Face*. Cambridge, Mass.: Harvard University Press.

-----. 1994. The Dewey Lecture. *Journal of Philosophy Philosophy* 91: 445-517.

-----. 2002. A Problem Above Reference. In M.J. Loux (ed). *Metaphysics: Contemporary Readings*. London: Taylor and Francis. pp 496-524.

Rabinow, P. and W.M. Sullivan (eds). 1987. *Interpretive Social Science*. Berkeley: University of California Press.

Radcliffe-Brown, A.R. 1922. *The Andaman Islanders*. (1964 Ed). New York: Free Press.

-----. 1952. *Structure and Function in Primitive Society*. London: Cohen & West.

-----. 1957. *A Natural Science of Society*. (Original 1948). New York: Free Press.

Radhakrishnan, S. 1932. *An Idealist View of Life*. London: Allen & Unwin.

-----. 1933. *Indian Philosophy. Vol. One*. London: Allen & Unwin.

Rand, A. 1964. *The Virtue of Selfishness: A New Concept of Egoism*. New York: New American History.

Rappaport, R. 1979. *Ecology, Meaning and Religion*. Berkeley, Ca.: North Atlantic Books.

Reyna, S.P. 1994. Literary Anthropology and the Case Against Science. *Man* 29/3: 555-581.

Richards, P. 1985. *Indigenous Agricultural Revolution: Ecology and Food Production in West Africa*. London: Hutchinson.

Richards, R.J. 2008. *A Tragic Sense of Life: Ernst Haeckel and the Struggle Over Evolutionary Thought*. Chicago: University of Chicago Press.

Richardson, J. 2004. *Nietzsche's New Darwinism*. Oxford: Oxford University Press.

Richardson, J. and B. Leiter (eds). 2001. *Nietzsche*. Oxford: Oxford University Press.

Ridley, M. 2010. *The Rational Optimists: How Prosperity Evolves*. London: Fourth Estate.

Robbins, B. 2011. Enchantment? No Thank You. In G. Levine (ed). *The Joy of Secularism*. Princeton: Princeton University Press. pp 74-94.

Rorty, R. 1980. *Philosophy and the Mirror of Nature*. Oxford: Blackwell.

-----. 1982. *Consequences of Pragmatism*. Minneapolis: University of Minnesota Press.

-----. 1989. *Contingency, Irony and Solidarity*. Cambridge: Cambridge University Press.

-----. 1991. *Objectivity, Relativism and Truth*. Cambridge: Cambridge University Press.

-----. 1998. *Truth and Progress*. Cambridge: Cambridge University Press.

Rose, H. and S. (eds). 2000. *Alas! Poor Darwin: Arguments Against Evolutionary Psychology*. London: Cape.

Rose, S. 1997. *Lifelines: Biology, Freedom and Determinism*. London: Penguin Books

-----. 2005. *The 21st Century Brain*. London: Cape.

Rose, S., L.J. Kamin and R.C. Lewontin. 1984. *Not in our Genes: Biology, Ideology and Human Nature*. London: Penguin Books.

Rotzer, F. 1995. *Conversations with French Philosophers*. (Original 1980). New Jersey: Humanities Press.

Rousseau, J.J. 1997. *The Discourses and Other Early Political Writings*. (Original 1751-1755). Cambridge: Cambridge University Press.

Roy, M.N. 1940. *Materialism: An Outline of the History of Scientific Thought*. (1982 Ed). Delhi: Ajanta Books.

Runciman, W.G. 2009. *The Theory of Cultural and Social Selection*. Cambridge: Cambridge University Press.

Rupp, R. 2005. *Four Elements: Water, Air, Fire, Earth*. London: Profile Books.

Russell, B. 1946. *History of Western Philosophy*. London: Allen & Unwin.

Ryan, A. et al. 1992. *After the End of History*. London: Collins & Brown.

Sahlins, M. 1976. *Culture and Practical Reason*. Chicago: University of Chicago Press.

-----. 1977. *The Use and Abuse of Biology*. London: Tavistock.

Sayer, A. 2000. *Realism and Social Science*. London: Sage.

Sayers, S. 1985. *Reality and Reason: Dialectic and the Theory of Knowledge*. Oxford: Blackwell.

-----. 1996. Engels and Materialism. In C.J. Arthur (ed). *Engels Today*. London: MacMillan.

Schmidt, A. 1971. *The Concept of Nature in Marx*. London: New Left Books.

Schmidt, L.K. 2006. *Understanding Hermeneutics*. Stocksfield: Acumen.

Schopenhauer, A. 1819. *The World as Will and Representation*. (1967 Ed). New York: Dover.

Schutz, A. 1967. *The Phenomenology of the Social World*. (Original 1932). Evanston, Ill.: Northwestern University Press.

Scott, J.C. 2009. *The Art of Not Being Governed*. New Haven: Yale University Press.

Scott, M.W. 2013. The Anthropology of Ontology (Religious Science?). *Royal Anthrop. Inst.* 19: 859-872.

Scruton, R. 2015. *Fools, Frauds and Firebrands: Thinkers of the New Left*. London: Bloomsbury.

Searle, J. 1999. *Mind, Language and Society: Doing Philosophy in the Real World*. London: Weidenfeld & Nicolson.

-----. 2009. Why Should You Believe It? *New York Rev. of Books* 24th September.

Seeland, K. (ed). 1997. *Nature is Culture: Indigenous Knowledge and Socio-Cultural Aspects of Trees*. London: Intermediate Technology Group.

Sellars, R.W. 1922. *Evolutionary Naturalism*. Chicago: Open Court.

Sen, A. 2006. *Identity and Violence: The Illusion of Destiny*. London: Allen Lane.

Sen, K.M. 1961. *Hinduism*. London: Penguin Books.

Sheehan, H. 1985. *Marxism and the Philosophy of Science*. New Jersey: Humanities Press.

Shweder, R.A. 1984. Anthropology's Romantic Rebellion Against the Enlightenment. In R.A. Shweder and R.A. Levine (eds). *Culture Theory: Essays in Mind, Self and Evolution*. Cambridge: Cambridge University Press. pp 27-66.

-----. 1986. Divergent Rationalities. In D.W. Fiske and R.A. Shweder (eds). *Metatheory in Social Science*. Chicago: University of Chicago Press.

Sim, S. (ed). 2005. *Routledge Companion to Postmodernism*. London: Routledge.

Simondon, G. 2012. *Being and Technology*. Edinburgh: Edinburgh University Press.

Skinner, B.F. 1953. *Science and Human Behaviour*. New York: Free Press.

Smith, A. 1970. *The Wealth of Nations*. (Original 1776). London: Penguin Books.

Smuts, J.C. 1926. *Holism and Evolution*. London: MacMillan.

Sokal, A. and J. Bricmont. 1999. *Intellectual Imposters: Postmodern Philosophers Abuse of Science*. London: Profile Books.

Sperber, D. 1996. *Explaining Culture: A Naturalistic Approach*. Oxford: Blackwell.

Steel, D. and F. Guala (eds). 2011. *The Philosophy of Social Science Reader*. London: Routledge.

Steward, J. 1955. *The Theory of Culture Change: The Methodology of Multilineal Evolution*. Urbana: University of Illinois Press.

Stich, S.P. 1999. *Deconstructing the Mind*. New York: Oxford University Press.

Stocking, G.W. 1992. *The Ethnographic Magic and Other Essays in the History of Anthropology*. Madison: University of Wisconsin Press.

-----. 1996. *After Tyler: British Social Anthropology 1888-1951*. London: Athlone Press.

Stollen, P. 1989. *The taste of Ethnographic Things: The Senses in Anthropology*. Philadelphia: University of Pennsylvania Press.

Stones, E. 1966. *An Introduction to Educational Psychology*. London: Methuen.

Strathern, M. 1988. *The Gender of the Gift*. Berkeley: University of California Press.

Symons, D. 1992. On the Use and Abuse of Darwinism in the Study of Human Behaviour *in* J.H. Barkow et al (eds) pp 137-157.

Tallis, R. 2011. *Aping Mind: Neuromania, Darwinitis and the Misrepresentation of Humanity*. Durham: Acumen.

-----. 2012. *In Defence of Wonder and Other Philosophical Reflections*. Durham: Acumen.

Tarde, G. 2000. *Social Laws: An Outline of Sociology*. Kitchener, Ontario: Batoche Press.

Taylor, C. 1985. *Philosophy and the Human Sciences*. Cambridge: Cambridge University Press.

Tester, K. 1991. *Animals and Society*. London: Routledge & Kegan Paul.

Thayer, H.S. 1981. *Meaning and Action: A Critical History of Pragmatism*. Indianapolis: Hackett Publications.

Thornhill, R. and C.T. Palmer. 2000. *The Natural History of Rape*. Cambridge, Mass.: MIT Press.

Timpanaro, S. 1975. *On Materialism*. London: New Left Books.

Todorov, T. 2009. *In Defence of the Enlightenment*. London: Atlantic Books.

Tooby, J. and L. Cosmides (eds). 1992. The Psychological Foundations of Culture, *in* J.H. Barkow et al. pp 19-36.

-----. 2006. Toward Mapping the Evolved Functional Organization of the Mind and Brain. In E. Sober (ed). *Conceptual Issues in Evolutionary Biology*. Cambridge, Mass.: MIT Press.

Trigg, R. 1980. *Reality at Risk*. Hemel Hempstead: Harvester.

Trigger, B.G. 1998. *Socio-Cultural Evolution*. Oxford: Blackwell.

Tudge, C. 2003. *So Shall We Reap*. London: Penguin Books.

-----. 2013. *Why Genes are not Selfish and People are Nice*. Edinburgh: Floris Books.

Turner, T. 1993. Anthropology and Multiculturalism. *Cultural Anthro*pology. 8: 411-429.

Tyler, S.A. 1986. Postmodern Ethnography: From Document of the Occult to Occult Document. In J. Clifford and G.E. Marcus (eds). *Writing Culture: The Poetics and Politics of Ethnography*. Berkeley: University of California Press. pp 122-140.

-----. 1991. A Post-Modern In-Stance. In L. Nencel and P. Pels (eds). *Constructing knowledge: Authority and Critique in Social Science*. London: Sage. pp 78-94.

Urry, J. 2003. *Global Complexity*. Cambridge: Polity Press.

Vygotsky, L. 1978. *Mind in Society: The Development of Higher Psychological Processes*. Cambridge, Mass.: Harvard University Press.

Wagner, R. 1981. *The Invention of Culture*. (Revised Ed). Chicago: University of Chicago Press.

Walsh, D.M. 2015. *Organism, Agency and Evolution*. Cambridge: Cambridge University Press.

Warren, W.P. 1975. *Roy Wood Sellars*. Boston: Twayne Publishers.

Watkins, J. 1968. Methodological Individualism and Social Tendencies. In M. Brodbeck (ed*). Readings in the Philosophy of the Social Sciences*. New York: MacMillan.

Webb, J., T. Schirato, and G. Danaher. 2002. *Understanding Bourdieu*. London: Sage.

West, C. 1989. *The American Evasion of Philosophy: A Genealogy of Pragmatism*. London: MacMillan.

Wheeler, H. (ed). 1973. *Beyond the Punitive Society*. London: Wildwood House.

Wheeler, H.M. 1928. *Emergent Evolution and the Development of Societies*. New York: Norton.

Wheen, F. 1999. *Karl Marx*. London: Fourth Estate.

White, L. 1949. *The Science of Culture*. New York: Grove Press.

Whitehead, A.N. 1920. *The Concept of Nature*. Cambridge: Cambridge University Press.

-----. 1929. *Process and Reality.* New York: Free Press.

-----. 1933. *Adventure of Ideas.* (1964 Ed). Cambridge: Cambridge University Press.

Whitehouse, H. (ed). 2001 *The Debated Mind.* Oxford: Berg.

Whitrow, G.J. 1975. *The Nature of Time.* London: Penguin Books.

Wilson, E.O. 1975. *Sociobiology: The New Synthesis.* Cambridge, Mass.: Harvard University Press.

-----. 1978. *On Human Nature.* (2004 Ed). Cambridge, Mass.: Harvard University Press.

-----. 1994. *Naturalist.* Washington: Warner Books.

-----. 1997. *In Search of Nature.* London: Penguin Books.

-----. 1998. *Consilience: The Unity of Knowledge.* London: Abacus.

Wittgenstein, L. 1922. *Tractatus Logico-Philosophicus.* London: Routledge & Kegan Paul.

Wolf, E. 1964. *Anthropology.* Englewood Cliffs: NJ Prentice Hall.

Wood, A.W. (ed). 1984. *Self and Nature in Kant's Philosophy.* Ithaca: Cornell University Press.

Wood, E.M. and J.B. Foster (eds). 1997. *In Defence of History: Marxism and the Postmodern Agenda.* New York: Monthly Review Press.

Woods, A. and T. Grant. 1995. *Reason in Revolt: Marxist Philosophy and Modern Science.* London: Wellred Publications.

Woolgar, S. 1986. On the Alleged Distinction Between Discourse and Practices. *Social Studies in Science* 16: 309-317.

Wrong, D. 1961. The Over-Socialized Conception of Man in Modern Sociology. *American Sociological Review* 26: 183-193.

Wulf, A. 2015. *The Invention of Nature: The Adventures of Alexander Von Humboldt, the Lost Hero of Science.* London: J. Murray.

Yolton, J.W. (ed). 1991. *The Blackwell Companion to the Enlightenment.* Oxford: Blackwell.

Zeitlyn, D. and R. Just. 2014. *Excursions in Realist Anthropology.* Newcastle: Cambridge Scholars Publishing.

Zerzan, J. 2008. *Twilight of the Machines.* Port Townsend: Feral House.

Zizek, S. 2004. *Organs Without Bodies: On Deleuze and Consequences.* London: Routledge.

Index